I0482664

The Water-Quality Monitoring Program for the Baltimore Reservoir System, 1981–2007—Description, Review and Evaluation, and Framework Integration for Enhanced Monitoring

By Michael T. Koterba, Marcus C. Waldron, and Tamara E.C. Kraus

Prepared in cooperation with the
City of Baltimore, Baltimore County, and Carroll County, Maryland

Scientific Investigations Report 2011–5101

U.S. Department of the Interior
U.S. Geological Survey

U.S. Department of the Interior
KEN SALAZAR, Secretary

U.S. Geological Survey
Marcia K. McNutt, Director

U.S. Geological Survey, Reston, Virginia: 2011

For more information on the USGS—the Federal source for science about the Earth, its natural and living resources, natural hazards, and the environment, visit http://www.usgs.gov or call 1-888-ASK-USGS

For an overview of USGS information products, including maps, imagery, and publications, visit http://www.usgs.gov/pubprod

To order this and other USGS information products, visit http://store.usgs.gov

Suggested citation:
Koterba, M.T., Waldron, M.C., and Kraus, T.E.C., 2011, The water-quality monitoring program for the Baltimore reservoir system, 1981–2007—Description, review and evaluation, and framework integration for enhanced monitoring: U.S. Geological Survey Scientific Investigations Report 2011–5101, 133 p.

Contents

Figures

Tables

Conversion Factors

Multiply	By	To obtain
Length		
inch (in.)	2.54	centimeter (cm)
inch (in.)	25.4	millimeter (mm)
foot (ft)	0.3048	meter (m)
mile (mi)	1.609	kilometer (km)
Area		
acre	4,047	square meter (m^2)
acre	0.4047	hectare (ha)
acre	0.004047	square kilometer (km^2)
square mile (mi^2)	259.0	hectare (ha)
square mile (mi^2)	2.590	square kilometer (km^2)
Volume		
gallon (gal)	3.785	liter (L)
gallon (gal)	0.003785	cubic meter (m^3)
million gallons (Mgal)	3,785	cubic meter (m^3)
billion gallons (Ggal)	3,785,000	cubic meter (m^3)
Flow rate		
cubic foot per second (ft^3/s)	0.02832	cubic meter per second (m^3/s)
cubic foot per second per square mile [(ft^3/s)/mi^2]	0.01093	cubic meter per second per square kilometer [(m^3/s)/km^2]
million gallons per day (Mgal/d)	0.04381	cubic meter per second (m^3/s)
million gallons per day per square mile [(Mgal/d)/mi^2]	1,461	cubic meter per day per square kilometer [(m^3/d)/km^2]
billion gallons per day (Ggal/d)	43.81	cubic meter per second (m^3/s)
billion gallons per day per square mile [(Ggal/d)/mi^2]	1,461,000	cubic meter per day per square kilometer [(m^3/d)/km^2]
Mass		
ounce, avoirdupois (oz)	28.35	gram (g)
pound, avoirdupois (lb)	0.4536	kilogram (kg)

Temperature in degrees Celsius (°C) may be converted to degrees Fahrenheit (°F) as follows:

$$°F = (1.8 \times °C) + 32$$

Specific conductance (conductivity) is given in micromhos per centimeter (micromhos/cm).

Concentrations of chemical constituents in water are given either in milligrams per liter (mg/L) or micrograms per liter (µg/L).

Water year for surface-water supply is the 12-month period October 1 through September 30. The water year is designated by the calendar year in which it ends and includes 9 of the 12 months.

Abbreviations

α	Accepted statistical significance level
BMC	Baltimore Metropolitan Council
BMP	Best Management Practice
CWA	Clean Water Act
DBP	Disinfection by-product
DPW	Department of Public Works, Baltimore, Maryland
DEPRM	Baltimore County Department of Environmental Protection and Sustainability (formerly Resource Management)
HAA	Haloacetic acid
ICPRB	Interstate Commission on the Potomac River Basin
MDE	Maryland Department of the Environment
NPDES	National Pollution Elimination Discharge System
p	Statistical power of test
RTG	Reservoir Technical Group
SDWA	Safe Drinking Water Act
SRBC	Susquehanna River Basin Commission
THM	Trihalomethane
RWMA	Reservoir Watershed Management Agreement
RWPC	Reservoir Watershed Protection Committee (earlier version of RTG)
RWPS	Reservoir Watershed Protection Subcommittee (earlier version of RWPC)
USEPA	U.S. Environmental Protection Agency
WWTP	Waste Water Treatment Plants

The Water-Quality Monitoring Program for the Baltimore Reservoir System, 1981–2007—Description, Review and Evaluation, and Framework Integration for Enhanced Monitoring

By Michael T. Koterba, Marcus C. Waldron, and Tamara E.C. Kraus

Abstract

The City of Baltimore, Maryland, and parts of five surrounding counties obtain their water from Loch Raven and Liberty Reservoirs. A third reservoir, Prettyboy, is used to resupply Loch Raven Reservoir. Management of the watershed conditions for each reservoir is a shared responsibility by agreement among City, County, and State jurisdictions. The most recent (2005) Baltimore Reservoir Watershed Management Agreement (RWMA) called for continued and improved water-quality monitoring in the reservoirs and selected watershed tributaries. The U.S. Geological Survey (USGS) conducted a retrospective review of the effectiveness of monitoring data obtained and analyzed by the RWMA jurisdictions from 1981 through 2007 to help identify possible improvements in the monitoring program to address RWMA water-quality concerns.

Long-term water-quality concerns include eutrophication and sedimentation in the reservoirs, and elevated concentrations of (a) nutrients (nitrogen and phosphorus) being transported from the major tributaries to the reservoirs, (b) iron and manganese released from reservoir bed sediments during periods of deep-water anoxia, (c) mercury in higher trophic order game fish in the reservoirs, and (d) bacteria in selected reservoir watershed tributaries. Emerging concerns include elevated concentrations of sodium, chloride, and disinfection by-products (DBPs) in the drinking water from both supply reservoirs. Climate change and variability also could be emerging concerns, affecting seasonal patterns, annual trends, and drought occurrence, which historically have led to declines in reservoir water quality.

Monitoring data increasingly have been used to support the development of water-quality models. The most recent (2006) modeling helped establish an annual sediment Total Maximum Daily Load to Loch Raven Reservoir, and instantaneous and 30-day moving average water-quality endpoints for chlorophyll-*a* (chl-*a*) and dissolved oxygen (DO) in Loch Raven and Prettyboy Reservoirs. Modelers cited limitations in data, including too few years with sufficient stormflow data, and (or) a lack of (readily available) data, for selected tributary and reservoir hydrodynamic, water-quality, and biotic conditions. Reservoir monitoring also is too infrequent to adequately address the above water-quality endpoints.

Monitoring data also have been effectively used to generally describe trophic states, changes in trophic state or conditions related to trophic state, and in selected cases, trends in water-quality or biotic parameters that reflect RWMA water-quality concerns. Limitations occur in the collection, aggregation, analyses, and (or) archival of monitoring data in relation to most RWMA water-quality concerns.

Trophic, including eutrophic, conditions have been broadly described for each reservoir in terms of phytoplankton production, and variations in production related to typical seasonal patterns in the concentration of DO, and hypoxic to anoxic conditions, where the latter have led to elevated concentrations of iron and manganese in reservoir and supply waters. Trend analyses for the period 1981–2004 have shown apparent declines in production (algal counts and possibly chl-*a*). The low frequency of phytoplankton data collection (monthly or bimonthly, depending on the reservoir), however, limits the development of a model to quantitatively describe and relate temporal variations in phytoplankton production including seasonal succession to changes in trophic states or other reservoir water-quality or biotic conditions.

Extensive monitoring for nutrients, which, in excessive amounts, cause eutrophic conditions, has been conducted in the watershed tributaries and reservoirs. Data analyses (1980–90s) have (a) identified seasonal patterns in concentrations, (b) characterized loads from (non)point sources, and (c) shown that different seasonal patterns and trends in nutrient concentrations occur between watershed tributaries and downstream reservoirs. A lack of data for total nitrogen and (or) available phosphorus limits direct comparisons of temporal or spatial variations in nutrient availability (comparable forms or ratios) between watershed tributaries and reservoirs.

Eutrophic conditions in the shallow water layer (30 feet in depth or less) in each reservoir have been assessed with four Carlson Trophic State Indices (TSIs)—derived from concentrations of chl-*a*, total phosphorus (TP), or DO, and (Secchi disc) transparency data. The frequency of eutrophic conditions for the entire period from 1982–2000 differed within each reservoir, and among the reservoirs, depending on which TSI index was used. The use of each index to compare trophic conditions among the reservoirs, however, possibly is biased because of the manner by which TSI data were collected, aggregated, or analyzed. In addition, no analyses of these indices were encountered that assessed possible trends in the frequency of eutrophic or mesotrophic conditions during this period.

Analyses of suspended-sediment data (1982–mid-90s) indicate that tributary concentrations and loads varied markedly within a year, and from year to year, but were clearly highest in wet years. Most sediment is carried by storm- as opposed to dry-weather (low) flow. Sediment transport has reduced reservoir capacity by 3 to 11 percent, and remains the major source of the TP load to the reservoirs. The role of this sediment as a source of available phosphorus (unmeasured) for phytoplankton production, however, has not been adequately addressed.

Manganese and iron are frequently monitored in water-supply intake waters during reservoir stratification and initial turnover. Elevated concentrations of these metals often occur at the supply intakes following their release from reservoir bed sediments under anoxic conditions, which can result from the decomposition of algal bloom residues. Monitoring in the reservoirs is too infrequent (monthly to bimonthly) to provide sufficient advanced warning of their occurrence at the intakes.

Elevated concentrations of mercury in game fish in the reservoirs are considered the end result of atmospheric deposition and beyond the control of RWMA jurisdictions. The submergence of terrestrial plants established on reservoir bed sediments exposed during droughts could enhance methyl-mercury production and biological uptake during reservoir recovery. However, this cannot be determined by conventional synoptic monitoring for mercury in game fish.

Fecal coliform bacteria have occurred at elevated counts in selected reservoir watershed tributaries, but counts in supply-reservoir intake waters consistently have been below the State recreational water-contact standard. Depending on results from synoptic surveys conducted by RWMA jurisdictions in the watersheds, the State could require routine monitoring of bacteria in the tributaries.

Among emerging concerns, trihalomethanes (THMs) and haloacetic acids (HAAs) are DBPs created by chlorination that are present in the drinking-water distribution systems of both supply reservoirs. Analysis of DBP data (2003–08) by the USGS indicates that the total concentrations of THMs and HAAs could exceed Federal standards under a pending rule change on approximately 19 percent and 40 percent of the sampling dates, respectively, at one or more monitoring stations in each water-distribution system. THM concentrations in drinking water varied seasonally, whereas HAA concentrations did not. There was little correlation between total concentrations of THMs and HAAs at a given monitoring station, or between monitoring-station concentrations of either DBP and total organic carbon (TOC) in intake waters. Monitoring of TOC alone will not identify intake waters associated with high concentrations of DBPs after chlorination.

In 2003, sodium and chloride concentrations at supply intakes were three-to-four-times greater than in the 1970s. Concentrations generally peaked during the winter months. Watershed and reservoir monitoring do not include the collection of sodium data. Monitoring also is too infrequent to provide either advanced warning of elevated sodium and chloride concentrations at the supply-reservoir intakes, or timely information on reductions in their concentrations if management activities are implemented to reduce road-salt use—the suspected source of the recent increases.

Projected changes combined with the inherent variability in climate in the Mid-Atlantic region indicate more intense storms with heavy precipitation and more frequent drought conditions. These changes imply increases in storm-borne contaminants (nutrient, sediment, salt, and bacterial loads), which could adversely affect reservoir water quality, particularly during recovery from drought conditions. Monitoring of stormflow does not appear to be adequate to address climate change and variability.

The 2007 Baltimore Reservoir System monitoring program could be improved in three major areas: (a) the monitoring design framework, (b) the temporal and spatial resolution of water-quality assessments in the major tributaries and reservoirs, and (c) the management and archival of data. Improvements in the framework design could include adoption of a quantitative phytoplankton model, such as the Phytoplankton Ecology Group model. Such models describe intra-seasonal, seasonal, and annual variations in phytoplankton abundance and succession. The model data can be analyzed in relation to temporal variations in nutrients or TSIs. The characterization of these biotic and water-quality conditions could be evaluated in relation to temporal variations in climate by the collection of climatic and water-quality data that reflect the full range in tributary flows and reservoir hydrodynamics within a year and from year to year. The minimal monitoring data would include daily temperature (mean), daily precipitation (total and type), continuous or partial records of streamflows depending on the type of tributary monitoring station, and daily water levels, withdrawals, and releases from each reservoir. To aid in this evaluation, the monitoring framework could incorporate the routine use of statistical and modeling methods to help define, aggregate, analyze, and interpret data.

Improvements in spatial and temporal assessments of water-quality conditions could be realized with two major and selected minor modifications to historical monitoring. First, to quantify water-quality conditions for the full range in tributary flows in the reservoir watersheds, sampling could include 3 to 15 pre-defined high (or storm-) flows per year at each of seven stations—three historical stations in each of the two supply reservoirs and one new station on a tributary to Prettyboy Reservoir. Pre-defined base-flow conditions could be sampled at each station on a monthly fixed time interval. Second, two fixed-station continuous monitors could be established in each reservoir to provide daily 5-foot-depth-increment profiles for selected parameters—water temperature, DO, pH, specific conductance, chl-*a*, turbidity, and depth of measurement. Data from these monitors could be transmitted to water-treatment staff to provide advanced warning of potential problems with supply intake waters.

A comprehensive quality-assurance program and plan (QAPP) with clear lines of responsibility could help ensure collection of the correct type and quality of data. The QAPP would include the following: (a) clear and concise definitions of the data and data-quality requirements for each water-quality concern; (b) field and laboratory methods and analytical procedures to obtain and provide the required data; (c) procedures to archive, clearly remark, and qualify data, including quality-assurance and control data; (d) procedures to routinely evaluate collected data in relation to data requirements; and (e) procedures to modify and document changes in field and laboratory methods.

Introduction

The City of Baltimore, Maryland (hereafter referred to as the City) supplies drinking water obtained from three reservoirs to approximately 1.8 million people in the City and parts of five Maryland Counties (Anne Arundel, Baltimore, Carroll, Harford, and Howard). Contributing watersheds to these reservoirs are primarily located outside the City in two Maryland counties (Baltimore and Carroll). The City is primarily responsible for managing and monitoring the reservoirs, monitoring in the major watershed tributaries, and assessing reservoir and major tributary conditions that affect the quality of drinking water. As the reservoir watersheds lie largely outside the jurisdiction of the City, however, managing and assessing reservoir-watershed conditions that could affect reservoir water quality is shared by City, County, and State governments. This shared responsibility is outlined in a voluntary Reservoir Watershed Management Agreement (RWMA) and related Reservoir Watershed Action Strategy (RWAS).

Implementation of the RWMA and RWAS involves the Baltimore Metropolitan Council (BMC), which provides staff for RWMA coordination. Management of the RWMA is conducted by a Watershed Protection Committee (WPC), which informs the BMC (Management Committee) of ongoing work. The WPC also provides policy guidance to a Reservoir Technical Group (RTG), and reviews their technical work. The RTG, a professional advisory body, is responsible for guiding day-to-day operations of the RWMA under the RWAS. It also provides technical advice, assistance, and recommendations to the WPC and RWMA signatories or their designees.

The most recent (2005) RWMA and RWAS reflect knowledge gained in part from routine water-quality monitoring in the reservoirs and selected reservoir watershed tributaries, which began in the early 1980s. The resulting monitoring data have served a wide range of purposes. For example, data routinely collected by drinking-water purveyors, primarily on raw water obtained through intakes in each water-supply reservoir, coupled with knowledge gained from their long-term monitoring and treatment of reservoir waters, helps guide daily decisions on which intakes to use to withdraw water from the reservoirs in order to provide suitable potable water at reduced costs. Data obtained from routine monitoring in the reservoirs are used for periodic assessments of reservoir water quality in relation to designated recreational uses (water-contact activities such as fishing and non-motorized boating, where permitted) and in relation to the general ecological health or trophic state of each reservoir. Routine monitoring in the reservoirs and selected watershed tributaries provides data to periodically characterize states, changes, or trends in water quality in the reservoirs and tributaries, and target management and restoration activities in the watersheds. Monitoring data also have aided in the development of watershed-reservoir models, which are used to guide management strategies to improve water quality in the watersheds tributaries and reservoirs.

As with most long-term monitoring efforts, the City and its RWMA partners recognize that the design and scope of the monitoring program require periodic evaluation. The purpose of this report is to aid the RWMA partners, and, in particular, the RTG, in an evaluation of the monitoring program as follows:

a) To describe the long-term and emerging monitoring-related RWMA water-quality concerns for the Baltimore reservoir system;

b) To evaluate the historical (1981–2007) and current (as of 2007) monitoring program in relation to its ability to provide suitable, relevant, and technically sound data to characterize water-quality conditions directly related to long-term and emerging water-quality concerns and in relation to expressed 2005 RWMA goals and action strategies; and

c) To provide a framework to identify continuing and additional monitoring that could enhance the ability of the RWMA partners to address specific water-quality concerns.

The scope of this report focuses on monitoring that was conducted either in the reservoirs or on selected major tributaries (subbasins) of the reservoir watersheds chiefly from the early 1980s (1981 or 1982, depending on the water-quality parameter) through 2007. The scope of this report also is a retrospective by nature, in that the review and evaluation are conducted mainly on the basis of an examination of dozens of historical investigative and technical reports produced through 2007, which discussed the production, analysis, or utilization of monitoring data, described findings, and possibly recommended modifications to improve the monitoring program.

The reports used in this retrospective review and evaluation were obtained during interviews and (or) by follow-up requests to agencies within or contracted under 2005 or past reservoir agreements and action strategies to provide data and (or) information relative to the Baltimore Reservoir Drinking-Water System. The reports include internal as well as published documents produced over several decades from a variety of agencies and organizations. As a result, the historical documents differed in the level of technical and scientific analysis, and in the manner and form in which monitoring data or interpretive analysis were reported and described (tables or figures). Limitations in the former are noted in this report where applicable. Modifications to the latter, where presented in this report for illustrative purposes, were minimal, and were used to improve visual quality and (or) maintain consistency in the names of reservoirs, reservoir watersheds, monitoring stations, or other place names used throughout this retrospective report.

Baltimore Drinking-Water Reservoir System

Drinking water for the City and all or parts of five Maryland counties is supplied by three surface-water reservoirs—Liberty, Loch Raven, and Prettyboy—and their contributing watersheds, which are entirely located in the Piedmont Physiographic Province in central Maryland (fig. 1). Water for consumptive use is withdrawn at intakes located in two of the reservoirs—Liberty and Loch Raven, which hereafter are collectively referred to as the water-supply reservoirs. The third reservoir, Prettyboy, is mainly used to provide additional storage and to re-supply the Loch Raven Reservoir. In addition, and generally during drought conditions, supplemental water supplies are obtained from the Susquehanna River upstream of Conowingo Dam, which is located approximately 45 mi (miles) northeast of the City (fig. 1).

Watershed and Reservoir Characteristics

Liberty Reservoir watershed covers 164 mi^2 (square miles; fig. 1, table 1), mainly in Carroll County and partly in Baltimore County, Maryland. Major land uses in the watershed are agriculture (43 percent), forest (32 percent), and developed land (22 percent; Maryland Department of Planning, 2000a; Winfield and Sakai, 2003). Agricultural lands are mainly cropland and pasture (37 percent and 6 percent, respectively). Developed lands include major transportation corridors and areas with predominantly industrial, commercial, and (or) residential infrastructure. Surface water to the reservoir is primarily supplied by the North Branch Patapsco River. Reservoir property covers 9,200 acres (table 1)—or 9 percent of the total watershed area—of which 3,100 acres is open water at reservoir capacity, estimated to be 37.7 Ggal (billion gallons) in 2001.

Loch Raven Reservoir watershed, excluding the Prettyboy Reservoir and watershed (fig. 1), covers 223 mi^2 (table 1), mostly in Baltimore and Carroll Counties, with small parts in Harford County, Maryland and York County, Pennsylvania. Major land uses are forest (38 percent), agriculture (27 percent), developed (21 percent), and mixed open (15 percent) (Maryland Department of Planning, 2000b; Maryland Department of the Environment, 2004a). Agricultural lands are mainly pasture and cropland (17 percent and 10 percent, respectively). Surface water to the reservoir is supplied primarily by the Gunpowder River. Reservoir property covers 8,000 acres—or 5.7 percent of the total watershed area—of which about 2,400 acres is open water at reservoir capacity, estimated to be 19.1 Ggal as of 1998.

Prettyboy Reservoir watershed covers 80 mi^2 (table 1), mostly in Baltimore and Carroll Counties in Maryland, with a small part in York County, Pennsylvania (fig. 1). Major land uses in the Maryland part of the watershed include agriculture (50 percent), forest (38 percent), and developed (13 percent) lands (Baltimore County Department of Environmental Protection and Resource Management, 2008). Agricultural lands are primarily cropland and pasture (39 percent and 11 percent, respectively). Surface water to the reservoir is supplied primarily by the Gunpowder River. Reservoir property covers 7,380 acres—or 14.3 percent of the total watershed area (table 1)—of which 1,500 acres is open water at reservoir capacity, estimated to be 18.4 Ggal in 1998.

Reservoir Watershed Management and Reservoir Operation

The City owns, and through its Department of Public Works (DPW), operates the three reservoirs to provide treated drinking water from the Liberty and Loch Raven water-supply reservoirs to approximately 1.8 million residents of the City and parts of five adjacent counties—Anne Arundel, Baltimore, Carroll, Harford, and Howard. Exclusive rights to surface water in the reservoirs and the Maryland part of their contributing watersheds have been granted to the City by the State legislature. However, only approximately 8 percent (table 1) of the total watershed area that drains into the three reservoirs actually is owned and under direct control of the City. Since the mid-1970s, the City DPW has been aided in its efforts to maintain the quality of water supplies by signatory City, County, and State organizations to a series of reservoir and watershed protection agreements and action strategies leading to the (2005) RWMA and RWAS.

The water-quality related goals of the 2005 RWMA for the program are as follows (Reservoir Watershed Management Agreement, 2005, p. 5–6):

a) To ensure the three reservoirs and their respective watersheds will continue to serve as:

1) Sources of high-quality raw water for the Baltimore metropolitan water-supply system; and

2) Areas where the surface waters will continue to support existing environmental, wildlife-habitat, and aesthetic purposes, as well as beneficial recreational uses.

b) To ensure that water quality in the three reservoirs and their tributaries consistently meet all applicable water-quality standards established by Federal and State regulations.

c) To ensure continued satisfactory water quality in the reservoirs themselves, by adopting the following specific technical goals:

1) Maintain existing water quality in the reservoirs and their tributaries, and reduce phosphorus, sediment, bacterial, sodium and chloride loadings to the reservoirs (and their tributaries) to acceptable levels[1], in order to:

(i) Eliminate existing, and prevent future, water-quality impairments, as defined under the Federal Clean Water Act (CWA), Section 303(d);

(ii) Prevent health and nuisance (taste and odor) conditions from developing in the treated water; and

(iii) Assist Baltimore City and Anne Arundel, Carroll, Harford, and Howard Counties (as water providers) to meet the Federal Safe Drinking Water Act (SDWA) requirements.

2) To improve the safety and security of the metropolitan water supply by reducing the risk of hazardous material contamination of the reservoir watersheds.

d) To commit program participants to promote certain types of land use and certain stewardship practices within the watershed that are intended to minimize the delivery of certain types of pollutants (including sediment and nutrients) to the three reservoirs.

The 2005 RWMA is accompanied by the 2005 RWAS (Reservoir Watershed Management Agreement Action Strategy, 2005). This strategy includes and encourages program participants to continue a multi-decadal effort to promote land use and stewardship practices within the watersheds that are intended to reduce the delivery of selected pollutants (for example, nutrients and sediment) to the reservoirs.

Whereas the 2005 RWMA and RWAS are designed primarily for management of the reservoir watersheds, the City manages and operates the reservoirs to provide drinking-water supplies. Drinking water from Liberty Reservoir is produced at the Ashburton treatment facility, and drinking water from Loch Raven Reservoir is produced at the Montebello treatment facility.

Water levels in the three reservoirs vary seasonally in response to climatic conditions and withdrawals for supplies. Summer seasonal drawdowns in water levels are a normal part of reservoir operations; however, recovery from high demand or climate stresses is slow, particularly in the case of Liberty Reservoir. For example, it can take several months or more for the reservoirs to recover after a substantial decrease in water levels (Valcik, 1975; Winfield and Sakai, 2003). Therefore, variations in water levels guide daily management decisions on withdrawals from each water-supply reservoir and releases of water from Prettyboy Reservoir. To reduce the duration and extent of drawdown in any reservoir, and particularly in Liberty Reservoir, the City employs what is officially referred to as their "firming program," which represents the documented procedure that utilizes water levels to govern reservoir withdrawals (Loch Raven and Liberty Reservoirs) or releases (Prettyboy Reservoir) to meet supply demands (Winfield and Sakai, 2003). This term will be used hereafter in this report.

Under the firming program, and assuming that all reservoirs have sufficient reserves, withdrawals for drinking water generally are made from both Liberty and Loch Raven Reservoirs. Daily withdrawals are incrementally reduced from Liberty Reservoir, however, as a function of seasonal demand and its water levels. For example, during the period of highest demand (generally June through September) and assuming all reservoirs are near capacity, withdrawals from Liberty Reservoir are typically 160 Mgal/d (million gallons per day) or more. Withdrawals from this reservoir are incrementally reduced, however, when the water level near the intakes in the lower part of this reservoir falls below 415 ft (feet) to as little as 60 Mgal/d if the water level falls below 370 ft. Increased withdrawals from Loch Raven Reservoir are used to make up the shortfall in demand. If demands result in water levels falling below approximately 236 ft near the intakes in the lower part of Loch Raven Reservoir, water is released from Prettyboy Reservoir, which is approximately 18 mi upstream on the Gunpowder River, to resupply Loch Raven Reservoir. The DPW also can release water from Prettyboy Reservoir as necessary during warmer low-flow periods to help maintain the aquatic habitat for stocked trout along the Gunpowder River between the two reservoirs.

The primary goal of the City in withdrawing water from either water-supply reservoir is to obtain the highest quality of raw water in order to minimize treatment costs (Winfield and Sakai, 2003). This is achieved by withdrawing water from one or more vertical intakes located at different depths at gatehouses in the middle (Loch Raven Reservoir only) and (or) at the lower end of each water-supply reservoir. City staffs at each reservoir treatment facility generally decide which intake(s) to use to withdraw raw water, and, if multiple intakes are used, the mixing ratio of intake waters. Their decisions are guided by routine (daily-to-weekly) monitoring of intake waters coupled with knowledge obtained from the long-term monitoring and treatment of reservoir waters.

During extended dry periods, water demands could result in continued declines in water levels in all three reservoirs.

---

[1] "Acceptable" is not explicitly defined in the agreement, but can be considered guided by elements (i), (ii), and (iii).

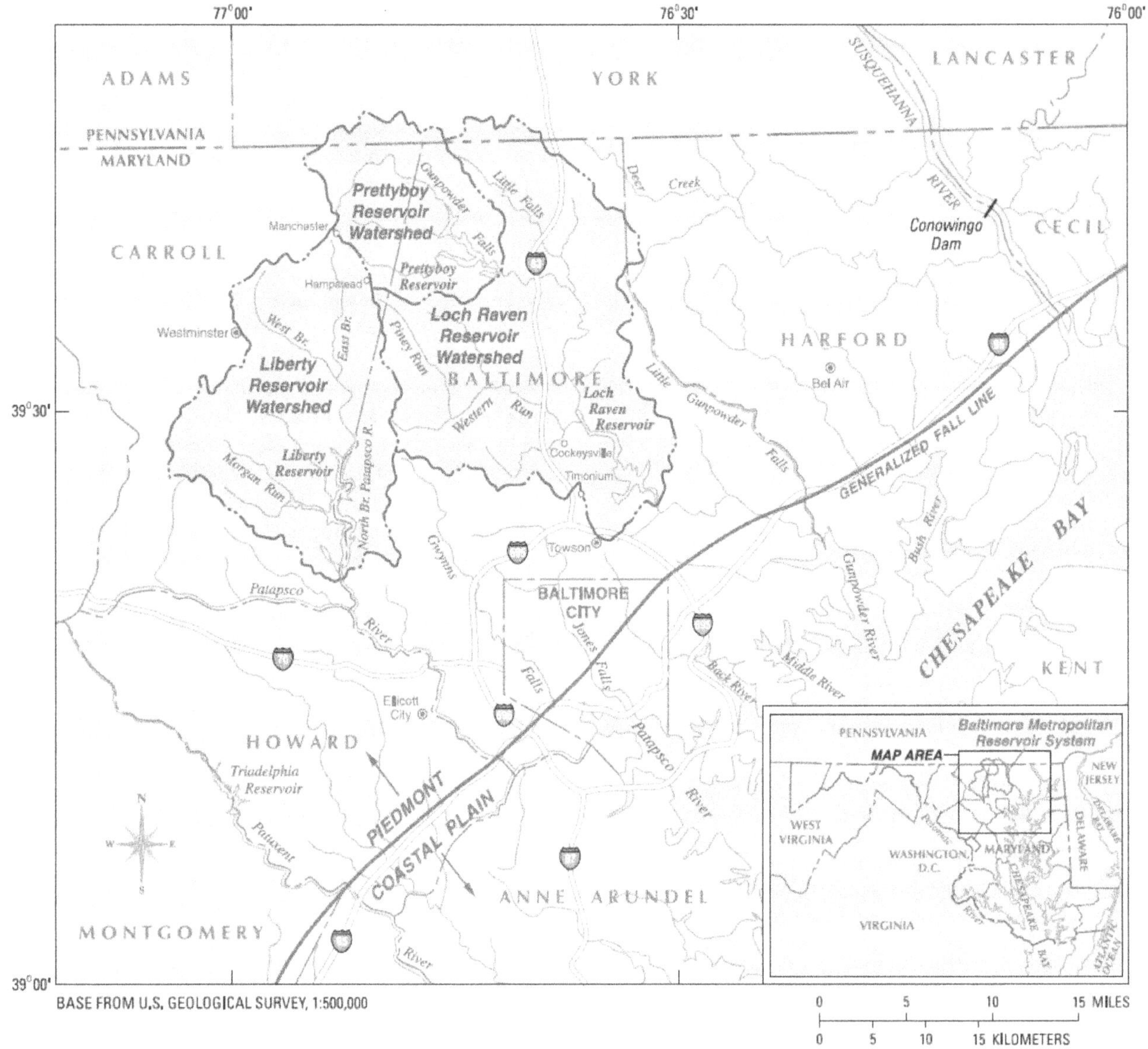

Figure 1.   Location of reservoirs and watersheds for the City of Baltimore, Maryland (modified from Baltimore Reservoir Technical Group, 2004).

The firming program generally has been able to circumvent this problem. Under extended withdrawals from all three reservoirs, water is released from Prettyboy Reservoir until the reservoir is at 50 percent of its capacity, whereupon the City can exercise its option to obtain water from the Susquehanna River at the Conowingo Dam (fig.1). Under an agreement with the Susquehanna River Basin Commission (2006), and dependent upon river flows to the Conowingo Dam, the City is permitted to pump from 64 to 240 Mgal/d on the basis of a 30-day average. The drainage area of the Susquehanna River Basin is 27,510 mi^2 above the dam (Susquehanna River Basin Commission, 2006), therefore, the low-flow limitation on City withdrawals typically only becomes a factor under prolonged regional droughts.

Generally it is the quality of the Susquehanna River water, and the added costs to the City to obtain, pump (transport), and treat this water, that limit its use. For example, during a severe drought in 2001–02, the City was able to obtain water of reasonably good quality from the Susquehanna River to help meet demands; nevertheless, major withdrawals and drawdowns ultimately occurred in all three reservoirs. During and upon recovery in 2003, however, the quality of water in the Baltimore Reservoirs declined in relation to selected water-quality conditions relative to pre-drought conditions (Baltimore Reservoir Technical Group, 2004). During a recent but less severe drought and recovery in 2005–06, the City also chose to use water from the Susquehanna River, but soon after the drought began, rather than withdrawing water solely

Table 1.   Reservoir and watershed characteristics for the City of Baltimore, Maryland, drinking-water supply system.

Reservoir/watershed	Characteristics[1]
Liberty	Area of watershed: 164 square miles
	Area of land owned by City: 9,200 acres or 14.4 square miles
	Storage capacity: Initial (1913) estimate, 40.0 billion gallons
	Storage capacity: Recent (2001) estimate, 37.7 billion gallons
	Length of shoreline at crest elevation: 82 miles
	Normal depth: 132.8 feet
	Flooded area at crest elevation: 3,106 acres
	Built: 1951–53, height 175 feet
Loch Raven	Area of watershed: 223 square miles (less Prettyboy watershed area)
	Area of land owned by City: 8,000 acres or 12.5 square miles
	Storage capacity: Initial (1913) estimate, 21.4 billion gallons
	Storage capacity: Recent (1997–98) estimate, 19.1 billion gallons
	Length of shoreline at crest elevation: 50 miles
	Normal depth: 76 feet
	Flooded area at crest elevation: 2,400 acres
	Built: 1912–14; crest raised: 1921–22, height 101 feet
Prettyboy	Area of watershed: 80 square miles
	Area of land owned by City: 7,380 acres or 11.4 square miles
	Storage capacity: Initial (1933) estimate, 19.9 billion gallons
	Storage capacity: Recent (1998) estimate, 18.4 billion gallons
	Length of shoreline at crest elevation: 46 miles
	Normal depth: 98.5 feet
	Flooded area at crest elevation 1,500 acres
	Built: 1933, height 155 feet

[1] Compiled from Ortt and others, 2000; Banks and LaMotte, 1999; Weisberg and others, 1985; and R. Ortt, Maryland Geological Survey, written commun., 2001.

from the reservoirs. The quality of river water in 2005–06 was notably poorer than the quality of river water in 2001–02, however, and use of this water was quickly discontinued (Michael Kohler, City of Baltimore, Department of Public Works, written commun., 2010).

As of 2007, the City firming program is under review (Michael Kohler, City of Baltimore, Department of Public Works, written commun., 2010). Decisions on when to begin reducing withdrawals from Liberty Reservoir, increase withdrawals from Loch Raven Reservoir, release water from Prettyboy Reservoir, or obtain water from the Conowingo Dam on the Susquehanna River, are still dependent upon the quality and available volumes of water. Comparing the costs to obtain, transport, and (or) treat each source of water for drinking water to provide the lowest-cost drinking water is becoming increasingly important to consider as part of the firming program. In addition, following the drought of 2001–02, the Susquehanna River Basin Commission (SRBC) initiated legal action to limit City withdrawals of water from the Conowingo Dam (Baltimore Reservoir Technical Group, 2004; Winfield and Sakai, 2003). The SRBC also is conducting an independent review of its basin management plan (Susquehanna River Basin Commission, 2006) because of increased demands for river water by upstream, in-lake, and downstream users, particularly during drought conditions.

Ultimately, how the City manages the reservoirs could possibly affect the quality of water in the reservoirs during their recovery following major droughts. During major droughts, the City maintains daily withdrawals of the best available quality of water for supplies to reduce treatment costs and limit consumer complaints about the quality of treated water. After recovery from droughts, there typically is a decline in the quality of reservoir waters, which could be exacerbated by the repeated removal of only the best available quality of water, as well as a considerable quantity of water, during drought conditions.

Overview of Water-Quality Concerns

Water-quality issues of concern to the RWMA partners relate primarily to impairments in designated recreational uses for the watershed tributaries, designated recreational and supply uses for the reservoirs, and ultimately, impairments in the quality of drinking water (fig. 2). Most of the water-quality issues of concern also are interrelated. The appearance of impairment conditions associated with one concern often can be concomitant or precede the appearance of additional impairments associated with other concerns.

For the purposes of discussion, the water-quality concerns identified by the 2005 RWMA can be grouped as long-term or emerging concerns. Long-term concerns include eutrophication and sedimentation in the reservoirs, and related to these concerns, elevated concentrations of (a) nutrients (nitrogen and phosphorus) being transported from the major tributaries to the reservoirs, (b) iron and manganese released from reservoir bed sediments during periods of deep-water anoxia, (c) mercury in higher trophic order game fish in the reservoirs, and (d) bacteria in selected reservoir watershed tributaries. Most of these concerns generally result, in part, from the initial creation of the reservoirs by the flooding of agrarian (mainly croplands) rather than undisturbed lands, and, in part, from subsequent decadal changes in land use and other human activities on the lands that remained. As a consequence of the former, the quality of reservoir waters was impaired in relation to most of these concerns from the start. As a result of the latter, there is evidence that the prolonged impairments in reservoir water and biotic quality have and will continue to make improvements to tributary and reservoir water quality in relation to these concerns a challenge.

Emerging water-quality concerns include elevated concentrations of disinfection by-products (DBPs), sodium, and chloride in the drinking water from both supply reservoirs. Climate change and variability also could be emerging concerns, affecting seasonal patterns, annual trends, and drought occurrence, which historically have led to declines in reservoir water quality. These emerging concerns reflect relatively recent (within the decade) changes in tributary or reservoir water quality, or in Federal standards related to drinking-water quality. Thus, the factor(s) responsible for these emerging water-quality concerns are not necessarily well known or understood.

Most long-term or emerging water-quality concerns have led to identified impairments in the designated uses for the reservoirs or their upstream tributaries. In turn, this has led to regulatory actions and concerns, which also influence monitoring and are described in detail later in this report. In addition, each of the long-term and emerging water-quality concerns is examined in detail later in this report in relation to the role that the long-term monitoring program has played in enabling the RTG to obtain the necessary data to understand and address each of these concerns.

Long-Term Water-Quality Concerns

Routine monitoring in the main tributaries in the reservoir watersheds and in-lake monitoring in the reservoirs only began in the early 1980s under the first formal voluntary agreement developed in the mid-to-late 1970s to manage the reservoir watersheds to maintain or improve the quality of water in the reservoirs. The main impetus for this agreement and monitoring was a host of reservoir and drinking-water impairments that occurred during a severe drought in the mid-to-late 1960s or during reservoir recovery in the early 1970s, or were apparent even earlier on the basis of limited monitoring conducted by City staff at the reservoir water-supply intakes. Examination of the possible causes of these impairments has a direct bearing on most RWMA long-term water-quality concerns and the initial design of the long-term monitoring program.

Amatayakul and others (1978), Amatayakul, Defries, and others (1978), Ortt and others (2000), Valcik (1975), and others describe the early Baltimore reservoir water-supply system from its inception to the creation of the first voluntary reservoir watershed protection agreement in the mid-1970s. The Baltimore reservoir system was created in the early through mid 20th century in a largely agrarian (cropland) landscape, which was mostly beyond the jurisdictional control of the City. Following creation of Loch Raven Reservoir (table 1, 1921–22), and up until World War II, most remaining croplands underwent conversion to pasture or woodlands. During World War II and thereafter, however, urban development (residential, commercial, and industrial) occurred in and near the reservoir watersheds, largely outward from the City along an arc on the southwest edge of the Liberty and Loch Raven Reservoir watersheds, and particularly in areas such as Towson, Maryland (fig. 1), and further north of Towson in Cockeysville and Timonium, Maryland. By the early 1970s, similar growth had occurred in reservoir watershed towns—for example, Hampstead and Manchester, Maryland, which are located along the divide between the Liberty Reservoir and Prettyboy Reservoir watersheds. Expansion of these areas continued through the 1980s and 1990s.

Stewart and others (2005) also noted that most of the aforementioned changes in land use occurred in the absence of environmental planning to protect tributary and reservoir water quality. For example, they estimated that 85 percent of all development in Baltimore County, which includes parts of all three reservoir watersheds (fig. 1), occurred before institutional programs were in place to mitigate potential adverse environmental impacts on streams or reservoirs.

Thus, by the year 2000, the landscape in each contributing reservoir watershed was no longer cropland, but a composite of forest, agricultural, and developed lands. Although land-use composition differs among reservoir watersheds, as of 2000, no watershed contained more than approximately 30–40 percent forest cover (see **Watershed and Reservoir Characteristics**, *this report*). Given the initial and subsequent

Watershed tributaries
- **Impaired biota, habitat, and recreational use**
  - Eutrophication in lower parts near reservoirs
  - Streambed-bank erosion
  - Elevated bacteria

Water treatment plant
- **Impaired drinking-water quality and quantity**
  - Increased treatment costs
  - Odor, color, taste
  - Elevated sodium and chloride (deicing salts)
  - Elevated disinfection byproducts from carbon source(s)
  - Decreased water supply

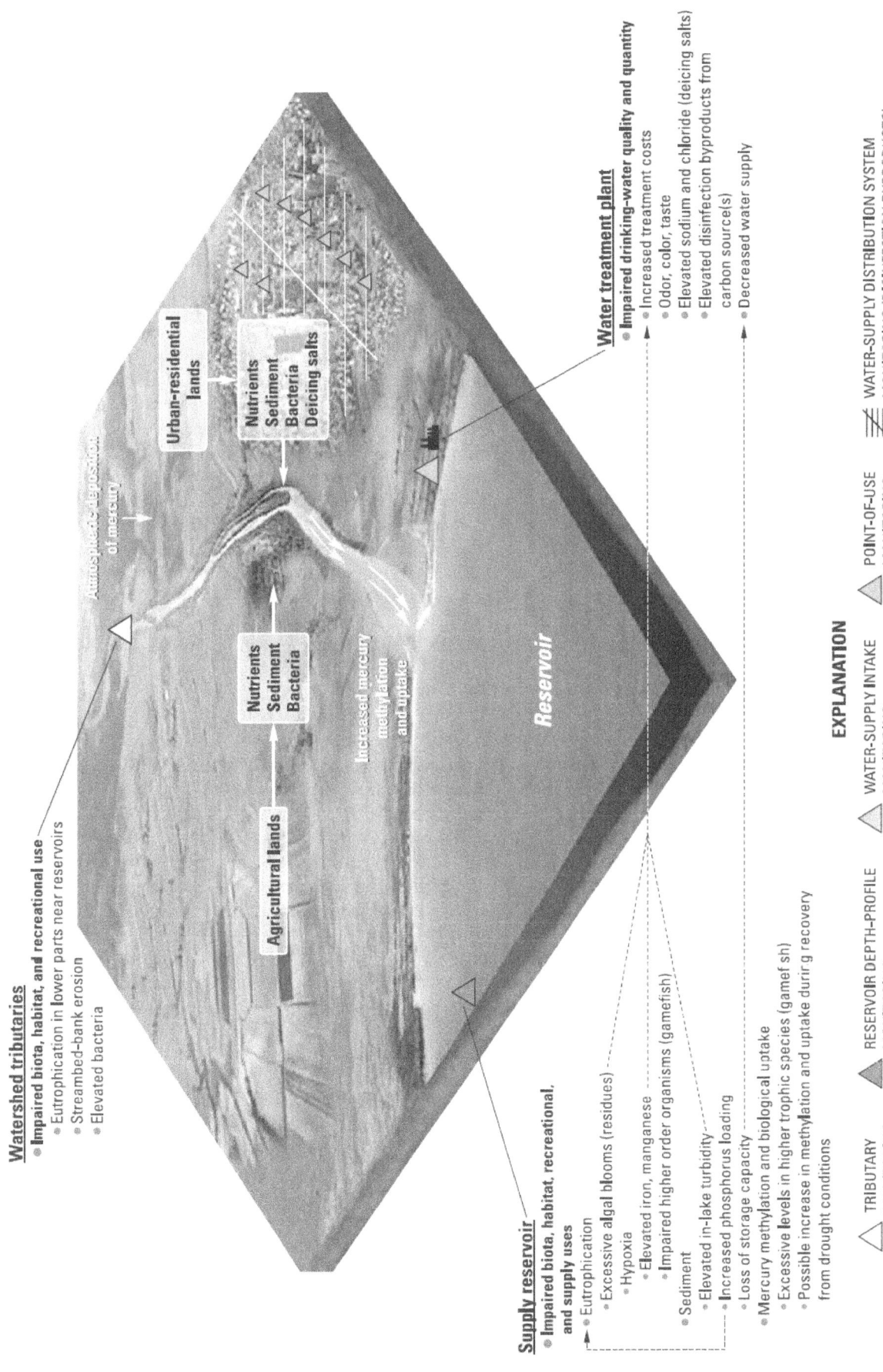

Urban-residential lands

Nutrients
Sediment
Bacteria
Deicing salts

Atmospheric deposition of mercury

Nutrients
Sediment
Bacteria

Increased mercury methylation and uptake

Agricultural lands

Reservoir

EXPLANATION

△ TRIBUTARY MONITORING

▲ RESERVOIR DEPTH-PROFILE MONITORING

△ WATER-SUPPLY INTAKE MONITORING

△ POINT-OF-USE MONITORING

⇇ WATER-SUPPLY DISTRIBUTION SYSTEM MONITORING (DISINFECTION BYPRODUCTS)

Supply reservoir
- **Impaired biota, habitat, recreational, and supply uses**
  - Eutrophication
    - Excessive algal blooms (residues)
    - Hypoxia
      - Elevated iron, manganese
      - Impaired higher order organisms (gamefish)
    - Sediment
      - Elevated in-lake turbidity
    - Increased phosphorus loading
    - Loss of storage capacity
  - Mercury methylation and biological uptake
    - Excessive levels in higher trophic species (gamefish)
    - Possible increase in methylation and uptake during recovery from drought conditions

Figure 2. Water-quality concerns related to impairments in watershed tributary, reservoir, or drinking-water quality and long-term monitoring in the Baltimore Reservoir System (watershed graphic modified from The H. John Heinz III Center for Science, Economics, and the Environment, 2008).

changes in land use, most long-term water-quality issues of concern to the RWMA partners likely arose when the reservoirs were created, or evolved during development that occurred before institutional measures were in place to reduce its adverse effects on the quality of water and biota in the reservoirs.

Large-scale monitoring in the Baltimore Reservoir System began long after most changes in land use, and largely evolved in response to reservoir and drinking-water quality impairments that arose with the severe drought and recovery that occurred in the late 1960s and early 1970s. Initial studies of the drought and recovery identified the watersheds as the ultimate sources of the pollutants that cause most long-term water-quality impairments (Valcik, 1975; Amatayakul and others, 1978; Amatayakul, Defries, and others, 1978). For example, in relation to the major phytoplankton blooms that occurred during the 1960s and 1970s, Valcik (1975) summarized several 1970 sanitary surveys conducted by the Maryland Department of Health and Mental Hygiene for each Baltimore reservoir watershed. He noted that poor agricultural practices—for example, possible overuse of chemical fertilizers and poor management of livestock and livestock manure—likely contributed to excessive nutrient pollution. Industrial and domestic wastewater effluents also were cited as possible contributors, in particular, the wastewater treatment plant discharges into Loch Raven Reservoir tributaries from Hampstead and Manchester, which began operations in 1970 and 1971, respectively. He also alluded to the high concentrations of nitrate found in groundwater in areas associated with these and other small towns and industries in Carroll and Baltimore Counties as sources of concern.

Further studies in the reservoir watersheds by the Baltimore City Department of Public Works (DPW) (1992, 1996, 2000, 2001) and others from the 1980s through the mid-1990s utilized monitoring data from the major tributaries to characterize pollutant (for example, nutrient and suspended-sediment) concentrations, trends, and loads in and from the reservoir watersheds. In addition, tributary monitoring data were analyzed to identify patterns or relations between tributary and reservoir water quality. Beginning in the late 1980s, analyses of tributary monitoring data also helped enable characterizations and source-water assessments of the environmental state of the reservoir watersheds and their tributaries (**Appendix A**). These characterizations and assessments laid the foundation for subsequent (post-1990s) short-term and synoptic monitoring studies in each reservoir watershed to aid in (a) source-water assessments, (b) identification of nonpoint pollutant problems and source areas, and (c) development and implementation of targeted restoration strategies in each reservoir watershed, to improve tributary water and habitat quality and reduce pollutant loads (**Appendix A**). As of 2007, the long-term water-quality issues of concern to the RWMA in the watershed tributaries continue to be eutrophication and major algal blooms caused by excessive concentrations of nutrients (nitrogen and phosphorus), suspended sediment, and bacteria (fig. 2).

Problems with reservoir water quality due to sedimentation and eutrophication actually were apparent even before the major drought and recovery of the 1960–70s. Amatayakul and others (1978), Amatayakul, Defries, and others (1978), Ortt and others (2000), Valcik (1975), and others describe the early Baltimore reservoir water-supply system and water-quality concerns. In 1881, the first Loch Raven Dam was installed about 1,000 ft downstream of the present Loch Raven Dam. This initial dam impounded about 510 Mgal of water. Heavy sedimentation rates, however, prompted construction of the current Loch Raven Dam and reservoir upstream of this original dam in 1914 (table 1). In addition to sedimentation, other water-quality problems were evident soon after the raising of the Loch Raven Dam in the 1920s. Amatayakul, Defries, and others (1978) analyzed long-term monitoring data from the Loch Raven Montebello treatment facility, and found that total microscopic counts (TMCs, in numbers per milliliter), a reflection of phytoplankton productivity, increased and remained elevated for several years after the reservoir rose. They also found long-term positive exponential rates of increase in average annual TMCs and chloride concentrations from 1930–55, markers of water-quality degradation. Significant exponential rates of decline in mean annual turbidity, bacterial counts, and concentrations of iron and manganese, however, indicated some improvement in water quality during this period.

During the 1960–70s, Valcik (1975) described a marked decline in supply reservoir water quality that was associated with the severe drought and recovery during this period. From 1963–70, total algal counts initially were low (approximately 100 organisms per milliliter) at the Montebello treatment facility for Loch Raven Reservoir. No major water-treatment problems with phytoplankton were encountered. Despite drought conditions, the flows at Loch Raven Dam during this period were about 160 Mgal/d, and reservoir water detention time was estimated to be about 4 months. In the early 1970s, with reservoir recovery, algal counts from intakes near the dam began to markedly increase, as did visual sightings of algal blooms in the upper reservoir and headwaters (fig. 1, Western Run and Piney Run). Major algal blooms re-occurred in 1973, and, by mid-July 1974, serious algal-related problems occurred in the treatment of reservoir water. Attempts to utilize intakes from the lower reservoir to supply suitable-quality water at low cost failed. Problems were encountered in the initial chlorination, alum coagulation, and sedimentation processes, with the removal of coagulated algal-alum residues, and with residual chlorination—all of which considerably increased the cost of potable water.

Valcik (1975) also described the deterioration of water quality in Liberty Reservoir. Under drought conditions from 1963–70, chloride concentrations at the Ashburton treatment facility increased from about 5 mg/L (milligrams per liter) to 15 mg/L; nitrate concentrations, which were being measured during that period, also more than doubled from about 2.5 mg/L to 6 mg/L as nitrate. Total algal counts at the Ashburton treatment facility were about 200 organisms per milliliter, or double the counts found at the Montebello treatment facility. Streamflow at the dam at Liberty Reservoir averaged only about 85 Mgal/d, or about half the flow found at the dam at Loch Raven Reservoir during this period. Low outflows coupled with the large storage capacity of Liberty Reservoir (table 1), which is about double the capacity of Loch Raven Reservoir, led to an estimated water detention time of 13 months, or more than three times the detention time for Loch Raven Reservoir during the drought period. Major seasonal algal blooms occurred in the upper Liberty Reservoir in 1963, 1965, 1968, and 1969. Despite re-occurring blooms, water-treatment problems were not encountered at the Ashburton treatment facility. Valcik (1975) suggested that the large capacity of this reservoir, combined with the use of deep intakes at the lower end of the reservoir, reduced the impact of blooms on the treatment of water. In fact, it appears multiple intakes were used, and that intake waters were mixed (Winfield and Sakai, 2003). Selected use of intakes, however, did not entirely eliminate adverse treatment consequences. Diatom blooms preferentially occurred in the deeper and cooler reservoir waters in 1967 and 1968. Their occurrence affected treatment operations, and increased treatment costs, but to a lesser degree than the costs that are typically incurred when treating waters with algal residues.

Subsequent bioassay and water-quality analyses by Amatayakul and others (1978) and Amatayakul, Defries, and others (1978) demonstrated (a) that additions of phosphorus and nitrogen to bioassay samples resulted in the greatest increase in phytoplankton production, and (b) that additions of phosphorus alone resulted in greater increases in phytoplankton production than additions of nitrogen alone. From these results, they concluded that phosphorus was the limiting metabolic nutrient. Amatayakul and others (1978) and Amatayakul, Defries, and others (1978) also analyzed a few reservoir profile samples from Loch Raven Reservoir. On the basis of the results, they concluded that bottom sediments were not a likely source of the phosphorus for algal blooms.

Following the 1960s–70s drought-recovery period, the reservoirs continued to experience excessive algal blooms (Baltimore Reservoir Technical Group, 2004). For example, in 1981, a month-long blue-green algal bloom occurred in Loch Raven Reservoir that resulted in major odor and taste issues in treated water and over 1,800 customer complaints. Nevertheless, residential development continued in several rural areas serviced by wastewater treatment plants (WWTPs) in headwaters of the reservoir watersheds. Concerned about further nutrient enrichment in the reservoirs as a result of proposed expansions of WWTPs and contributions from agricultural lands, City, County, and State officials agreed in 1979 to try to protect the reservoirs. Subsequent formal agreements and action strategies followed and led to the 2005 agreement and action strategy (Reservoir Watershed Management Agreement, 2005; Reservoir Watershed Management Agreement Action Strategy, 2005) described earlier in this report.

Monitoring in the reservoirs under these agreements began in the early 1980s. The period of digitally available monitoring data relative to the age of the reservoirs (table 1), however, is relatively short, and the breadth of data is limited in scope. Nevertheless, considerable use has been made of these data to characterize the initial state of water quality in the reservoirs, identify possible seasonal and annual trends, and aid in the development of reservoir and watershed models that could help manage the reservoirs. The results of these efforts have been periodically summarized by the City and (or) the RTG, and are described in detail in relation to each long-term water-quality concern later in this report.

Emerging Water-Quality Concerns

Emerging concerns for the RWMA partners since 2000 are the occurrence of elevated concentrations of DBPs, sodium, and chloride in treated water from each water-supply reservoir. These concerns chiefly reflect the need to address (a) the recent (2006) U.S. Environmental Protection Agency (USEPA) Safe Drinking Water Act (SDWA) Stage 2 rule change that limits concentrations of DBPs at specific locations, rather than on average throughout the drinking-water-distribution system (U.S. Environmental Protection Agency, 2006), and (b) the recent (2003) USEPA SDWA health advisory on sodium intake for individuals on restricted low-salt intake diets (U.S. Environmental Protection Agency, 2003).

Under the (2006) USEPA SDWA Stage 2 rule change, the 30-day moving averages for the total concentrations of selected trihalomethanes (THMs) and haloacetic acids (HAAs) cannot exceed the Maximum Contaminant Levels (MCLs) of 80 µg/L (micrograms per liter) and 60 µg/L, respectively, at any monitoring station within the treated water distribution system of either water-supply reservoir. Preliminary studies indicate that the total concentrations of either THMs or HAAs occasionally could exceed their respective MCLs at one or more stations in either distribution system (Maryland Department of the Environment, 2004a; Winfield and Sakai, 2003).

Since the early 1970s, when sodium concentrations began to be routinely measured, concentrations of sodium have almost tripled in supply-intake water from Liberty Reservoir and almost quadrupled in supply-intake water from Loch Raven Reservoir (Baltimore Reservoir Technical Group, 2004). Concentrations generally have peaked during the winter months. The suspected source of the recent increases is the use of sodium chloride (road salt) as a deicing agent (Winfield and others, 2006).

Although not formally considered an emerging concern by the RWMA partners, droughts and recoveries from drought have led to adverse water-quality conditions in the reservoirs as indicated above. Recent projected changes in climate, combined with the inherent variability in climate in the Mid-Atlantic region could lead to an increase in the number of intense storms with heavy precipitation and an increase in the frequency of extended dry periods or drought conditions (Intergovernmental Panel on Climate Change, 2007; Maryland

Department of the Environment, 2008). These changes imply that potential increases in storm-borne contaminants (nutrient, sediment, salt, and bacterial loads), could adversely affect reservoir water quality, particularly during recovery from drought conditions. Thus, the effects of climate are included as an additional emerging concern to be considered in this retrospective review of the monitoring program.

Regulatory Concerns

Water-quality concerns for the RWMA partners mainly relate to their goals, which include ensuring that SDWA and CWA Section 303(d) compliance continues. As of 2007, and except for periodic problems associated with color, odor, and taste, reservoir drinking-water supplies have routinely met the SWDA criteria for public water supplies. Also as of 2007, most water-quality concerns have led to impairments governed under Section 303(d) of the CWA, which provide general water-quality criteria that prohibit pollution of waters of the State of Maryland by any material in amounts sufficient to create nuisance or other interferences in designated uses (Code of Maryland Regulations (COMAR), 26.08.02.08). As of June 2008 (COMAR 26.08.02.08J, K)[2], the Maryland Water Quality Standards Stream Segment Designation for the reservoirs and their watershed tributaries can be described as follows:

a) Loch Raven and Prettyboy Reservoir Watersheds: Use III-P, all waters above the dam at Loch Raven, and therefore in the Loch Raven and Prettyboy Reservoir watersheds, are to be suitable for nontidal cold water (growth and propagation of trout, and supporting self-sustaining trout and their associated food organisms), as well as, for Use I-P, which implies suitable for recreation (water contact sports, fishing, and other play and leisure activities where individuals could come in direct contact with water surface), protection of nontidal warm-water aquatic life (growth and propagation of fish, other than trout), other aquatic life, and wildlife, and public water (as well as agricultural and industrial) supply; and

b) Liberty Reservoir Watershed: Designated uses differ depending on water body, as follows:

1) Use I-P, Reservoir and above reservoir, and, except for tributaries designated uses III, III-P, or IV-P below, all tributaries to the West Branch and North Branch of the Patapsco River;

2) Use III, all tributaries to and the main branch of Roaring Run on North Branch Patapsco River, which implies they are to be suitable for nontidal cold water as well as Use I, which implies suitable

for recreation, protection of nontidal warm-water aquatic life, other aquatic life, and wildlife;

3) Use III-P, all tributaries and main branches of Beaver Run, Cooks Branch, East Branch Patapsco River, Keysers Run, Locust Run, Morgan Run, Norris Run on the West Branch Patapsco River, see a) above; and

4) Use IV-P, main stems of West Branch Patapsco River (main stem only) and North Branch Patapsco River and Cranberry Branch and its tributaries (near Westminster, Maryland), above Liberty Reservoir are designated Use IV-P, which implies suitable for recreational trout waters (capable of holding and supporting adult trout for seasonal stocking and put and take fishing) and all uses under I-P, see a) above.

The inability of selected waters to fully meet the above designated-use criteria has led to CWA 303(d) listings of the impairments to reservoir watershed tributaries and (or) reservoirs, and of the progress made (as of 2007) to address each impairment (table 2). The Loch Raven and Prettyboy Reservoirs first appeared on the State 1996 Section 303(d) list as impaired by nutrients (mainly phosphorus), selected metals, and sediment (Loch Raven only). Listings for polychlorinated biphenyls (Loch Raven only) and mercury in fish tissue, bacteria, and biological communities were added in 2002.

Total Maximum Daily Loads (TMDLs) and final decision rationales for TMDLs for nutrients (phosphorus) and sediment for Loch Raven Reservoir, and nutrients (phosphorus) for Prettyboy Reservoir were approved by the USEPA (Maryland Department of the Environment, 2006). The specified water-quality goal of the nutrient TMDLs is to reduce the occurrence of high chlorophyll-a (chl-a) concentrations, which reflect excessive algal blooms, and to maintain dissolved-oxygen (DO) concentrations at levels supporting designated uses, in Loch Raven and Prettyboy Reservoirs. The water-quality goal of the sediment TMDL (28,925 tons per year, with 27,715 tons per year from nonpoint sources) for Loch Raven Reservoir is to increase the useful life of the reservoir for water supply by storage preservation.

Water-quality analyses for heavy metals (notably, chromium and lead) and polychlorinated biphenyls in fish tissue (only Loch Raven Reservoir), were completed and submitted to the USEPA for these reservoirs in 2003 (Maryland Department of the Environment, 2003 a,b). Results indicated that only manganese and iron, which are naturally abundant in watershed and reservoir sediments, occurred at elevated concentrations. As of 2007, it was anticipated that both reservoirs would be de-listed for polychlorinated biphenyls and the heavy metals in question, including chromium, lead, iron, and manganese.

The TMDLs for mercury in fish were completed for both reservoirs in 2002 (Maryland Department of the Environment, 2002 a,b and 2004 b,c), and included evidence that atmospheric depositional sources largely beyond the jurisdictional

---

[2] As of 2007, from the Maryland Department of Environment, accessed March 15, 2011 at *http //www.mde.maryland.gov/programs/water/TMDL/ Integrated303dreports/Pages/303d.aspx*.

Table 2. Federal Clean Water Act Section 303(d) impairments for Prettyboy, Loch Raven, and Liberty Reservoirs[1].

[MDE, Maryland Department of the Environment; TMDL, Total Maximum Daily Load; USEPA, U.S. Environmental Protection Agency; As, arsenic; Cd, cadmium; Cr, chromium; Cu, copper; Ni, nickel; Pb, lead; Se, selenium; Zn, zinc]

Reservoir	Impairment category: constituent	Year listed	Comments
Liberty	Nutrients: phosphorus	1996	Not fully addressed as of 2007
	Sediments: suspended sediments	1996	Not fully addressed as of 2007
	Metals: chromium and lead	1996	MDE submitted water-quality analysis to USEPA on September 24, 2003 supporting delisting for chromium and lead
	Metals: methylmercury—fish tissue	2002	MDE submitted TMDL to USEPA on December 27, 2002, with analysis supporting deferment and indicating chief sources are from atmospheric deposition
	Bacteria—*Escherichia coli sp.*	2002	Additional data requested by MDE in 2003; as of 2007, MDE was analyzing these data
	Biological communities	2002	Not yet addressed
Loch Raven	Nutrients: phosphorus	1996	MDE submitted TMDL final decision rationale to USEPA on March 27, 2007
	Sediments: suspended sediments	1996	MDE submitted TMDL final decision rationale to USEPA on March 27, 2007
	Metals: As, Cd, Cr (total), Cu, Ni, Pb, and Se	1996	MDE submitted water-quality analysis to USEPA on September 24, 2003 supporting delisting for heavy metals
	Metals: methylmercury and polychlorintaed biphenyls (PCBs)—fish tissue	2002	MDE submitted TMDL to USEPA on December 27, 2002, with analysis supporting deferment and indicating chief sources of mercury are from atmospheric deposition; as of 2003, reservoir was delisted for PCBs
	Biological communities	2002, 2004	Not fully addressed as of 2007
Prettyboy	Nutrients: phosphorus	1996	MDE submitted TMDL final decision rationale to USEPA on March 27, 2007
	Metals: As, Cd, Cr (hexavalent), Cu, Ni, Pb, Se, and Zn	1996	MDE submitted water-quality analysis to USEPA on September 24, 2003 supporting delisting for heavy metals
	Metals: methylmercury—fish tissue	2002	MDE submitted TMDL to USEPA on December 27, 2002, with analysis supporting deferment and indicating chief sources are from atmospheric deposition
	Bacteria—*Escherichia coli sp.*	2002	Additional data requested by MDE in 2003; as of 2007, MDE was analyzing these data
	Biological communities	2002, 2004	Not fully addressed as of 2007

[1] Accessed December 29, 2010 at http://www.mde.maryland.gov/programs/water/TMDL/Integrated303dReports/Pages/303d.aspx.

control of the City accounted for most of the mercury input into each reservoir. As of 2004, it was anticipated that no further action on mercury in fish tissues would be required until such time as the USEPA completes a review of sources of atmospheric mercury.

Listings for bacteria (Prettyboy only) and biological communities are mainly for watershed tributaries. As of 2007, action on both of these items awaited analysis of data by the Maryland Department of the Environment (MDE).

Liberty Reservoir first appeared on Maryland's 1996 Section 303(d) list of water quality-limited segments as impaired by nutrients (phosphorus), sediment, and metals (table 2). The listings for mercury in fish tissues, bacteria (tributaries only), and biological communities was added in 2002.

As of 2007, impairments for nutrients and sediment for Liberty Reservoir have yet to be addressed. Water-quality analyses for metals (in particular, for chromium and lead) were completed for this reservoir in 2003 (Maryland Department of the Environment, 2003c), and, as in the case of Prettyboy and Loch Raven Reservoirs, indicated that only manganese and iron, which are naturally abundant in reservoir sediments, occurred at elevated concentrations. The TMDL to address mercury was completed in 2002 (Maryland Department of the Environment, 2002c), with deferment anticipated for reasons similar to those noted above for Loch Raven and Prettyboy Reservoirs. Listings for bacteria and biological communities are awaiting action pending the collection and analysis of data by MDE.

Addressing regulatory impairments for nutrients (phosphorus) and sediment in Loch Raven Reservoir and for nutrients (phosphorus) in Prettyboy Reservoir provides a basis for future monitoring requirements. The rationale presented and accepted by the USEPA (Maryland Department of the Environment, 2006) to control phosphorus is focused on chl-*a* as the water-quality endpoint for the phosphorus TMDL. The chl-*a* concentration endpoints selected for the Loch Raven and Prettyboy Reservoirs are as follows:

a) A maximum permissible instantaneous chl-*a* concentration of 30 µg/L in surface layers; and

b) A 30-day moving average chl-*a* concentration not to exceed 10 µg/L in surface layers.

Under the same rationale (Maryland Department of the Environment, 2006), the concentration of DO is the water-quality endpoint for proposed and accepted nontidal designated biotic and designated uses of the Loch Raven and Prettyboy Reservoirs, which are as follows:

a) Daily average of 5.0 mg/L throughout the reservoir water column during periods of complete and stable mixing;

b) Daily average of 6.0 mg/L in the mixed surface layers at all times; and

c) Hypoxia in the deep water layers is to be addressed on a case-by-case basis, taking into account the morphology, degree of stratification or mixing of stratified waters in surface layers during lake turnover or drawdown, and given that seasonal hypoxia likely occurs regularly in both reservoirs in the hypolimnetic layer.

It also is apparent in the presented rationale that to the extent to which these TMDLs control excessive algal blooms, there is expected to be a corresponding reduction in the severity of DO sags, anoxic conditions, and the release of metals such as manganese and iron, as well as phosphorus, from bottom sediments. It is not apparent in the documentation how surface layers, mixed surface layers, or the hypolimnetic layer are to be defined, which is discussed later in this report.

In conjunction with excessive nutrients, it is assumed that the sediment impairment to Loch Raven Reservoir also would be addressed by the phosphorus TMDL. The underlying assumption is that the bulk of phosphorus entering this reservoir is bound to sediment. If true, then any control strategy directed toward reducing total phosphorus entering this reservoir also would help reduce sediment and vice versa. The MDE has adopted this rationale in all reservoirs and impoundments where both nutrient and sediment impairments exist.

The Water-Quality Monitoring Program

Water-quality monitoring in the Baltimore Reservoir System can be described in general terms from the formal implementation of the program in 1981–82 through 2007, as well as its current (2007) state. The effectiveness of this program can be described in relation to (a) a review and evaluation of the overall quality of the database and data collected during this period, (b) broad-based modeling attempts to describe and relate water-quality conditions in the reservoir watershed tributaries to water-quality and biotic conditions in the reservoirs, and (c) the use of monitoring data and information derived from the analyses of these data to describe spatial and temporal variations in water-quality parameters associated with each of the long-term and emerging water-quality concerns.

Description

Since its inception in the early 1980s, the long-term core monitoring network for the Baltimore Reservoir System generally has consisted of 21 nonpoint source (tributary or pond) and point-source water-quality monitoring stations in the reservoir watersheds, and 12 in-lake monitoring stations in the three reservoirs (fig. 3 and **Appendix B**). Use of this network and the data it has provided has varied over time. These variations are described in terms of a historical perspective (early1980s through 1990s or early 2000s) and a current perspective (chiefly as of 2007, given information provided by RWMA partners).

Historical Perspective (1981–2007)

Initially (1980s–90s), data from the core monitoring network were obtained and used mainly to define reservoir trophic conditions and their relation to drinking-water supplies, and describe variations in tributary water quality (Baltimore City Department of Public Works, 1992). By the mid-1990s, monitoring data were being used to describe and directly relate variations in tributary water quality to reservoir trophic conditions (Baltimore City Department of Public Works, 1996). It became apparent, however, that long-term monitoring would be required to detect trends that reflected the following: (a) the effect of management actions taken to reduce nonpoint sources of pollutants on tributary loads, (b) the effect of changes in tributary loads on reservoir trophic conditions, and (c) the effect of changes in reservoir trophic conditions on the quality of drinking-water supplies (Baltimore City Department of Public Works, 2000, 2001). The difficulty in the establishment of direct relations among tributary and reservoir water-quality conditions has been attributed in part to large variations in tributary and reservoir water quality due to year-to-year and within-year variations in streamflow, and in part to variations in in-stream and in-lake processing of pollutants. Nevertheless, the RWMA goals and action strategies remained clear—to reduce nutrient and sediment loads, and address emerging drinking water-quality concerns (Reservoir Watershed Protection Committee, 2000).

To help reduce nutrient and sediment loads and address these concerns, analysis of the monitoring data shifted in the mid-1990s from the initial emphasis on characterizing reservoir water quality to focus on tributary water quality and biotic conditions, and their relation to human activities in the watersheds (Baltimore City Department of Public Works, 2000, 2001). Unfortunately, resource limitations also notably reduced storm monitoring after the mid-1990s, and it has been intermittent thereafter through 2007.[3] Thus, characterizing and comparing water-quality tributary conditions among all subbasins in a reservoir watershed from the early 1980s through 1990s was done with data obtained from monthly dry-weather-flow sampling, rather than data collected from annual storm- and dry-weather flows. Dry-weather-flow sampling, as its name implies, is performed at all tributary stations during generally low-to-possibly moderate, and likely wadeable, flow conditions. These conditions historically are referred to by the RTG as dry-weather flows, and the resultant data, as dry-weather-flow data. The same terminology is used in this report as these flows have not been routinely quantified since the mid-1990s.

Additional support in the development of watershed-restoration action strategies to address RWMA partner concerns has come from numerous short-term and synoptic monitoring efforts in each watershed (see **Appendix A** for details). Most of these efforts were conducted by the City, Baltimore or Carroll County, a State agency—for example, MDE or the Maryland Department of Natural Resources (MDDNR)—or as a collaborative effort. Most of these efforts were conducted on smaller representative areas in selected subbasins within each reservoir watershed, and at different times among the reservoir watersheds as resources became available. These short-term (single-year) and synoptic (low-flow) monitoring efforts, nevertheless, have aided the RWMA partners in the development of reservoir-watershed source-water assessments, characterization studies, and management plans with restoration strategies. The chief objectives of these efforts have been to help identify, prioritize, plan, implement, manage, and protect or restore subbasin conditions to reduce identified water-quality pollutant point and nonpoint tributary sources and loads, and to meet Federal requirements and State programs.

The reservoir-watershed source-water assessments were conducted to evaluate the safety of all public drinking-water systems. Subsequent to or concomitant with the source-water assessments, selected RWMA partners conducted watershed-characterization surveys, nutrient and biotic surveys, and stream-corridor and stability assessments, all of which were used to develop a watershed-restoration action strategy. The restoration action strategies have been used to direct resources towards restoration activities in each reservoir watershed.

All of the above assessments, surveys, or restoration action strategies largely were completed for the major part (Carroll County) of the Liberty Reservoir watershed by 2004, and the major part of the lower Loch Raven Reservoir watershed (Baltimore County part, below Prettyboy Reservoir) by 2005. The nutrient synoptic and stream-corridor surveys were completed for the Prettyboy Reservoir watershed in 2008, as were the watershed characterization and restoration action strategy.

Current Perspective (as of 2007)

The long-term monitoring network and strategy for the Baltimore reservoirs and their contributing watersheds were described in the latest recent Reservoir Watershed Management Agreement Action Strategy (2005)—a description which remains accurate as of 2007. The City DPW is responsible for conducting comprehensive water-quality monitoring at the watershed tributary stations and at the in-lake stations of the three reservoirs (fig. 3, and **Appendix B**, table B1), and for the analysis of the monitoring samples for contaminants of concern (**Appendix B**, tables B2 through B4). They also are responsible for periodically analyzing and summarizing water-quality data for the RTG and RWMA partners on the status and trends in watershed tributary and reservoir water quality, including calculating annual loadings of selected pollutants—mainly suspended sediment and total phosphorus. Baltimore County is primarily responsible for selected chemical and biological monitoring in the tributaries in its parts of the three reservoir watersheds, and reporting the results annually in the County's National Pollution Discharge Elimination System (NPDES) report to MDE. The U.S. Geological Survey (USGS) is mainly responsible for providing discharge data for the six tributary stations where sampling is conducted during selected storms as well as dry-weather flows.

As part of the 2005 Action Strategy, the RTG is responsible for initiating and overseeing the evaluation of reservoir monitoring programs, and determining the resources needed to develop and maintain an integrated, comprehensive monitoring program. The monitoring network as of 2007 will be examined in relation to the following abilities:

a) The ability to detect annual and long-term water-quality trends in the reservoirs and their contributing watersheds, emphasizing the reservoirs as a source of potable water and a habitat for desirable living resources;

b) The effectiveness of the monitoring network to support predictive tools (such as computer models) in helping to manage reservoir water quality;

c) The implementation of new technologies to improve the effectiveness of reservoir watershed-management efforts;

d) The ability to link various types of pollutant sources to 2000 land use and land cover in the watersheds; and

---

[3] William Stack, Baltimore City Department of Public Works, written commun., 2008.

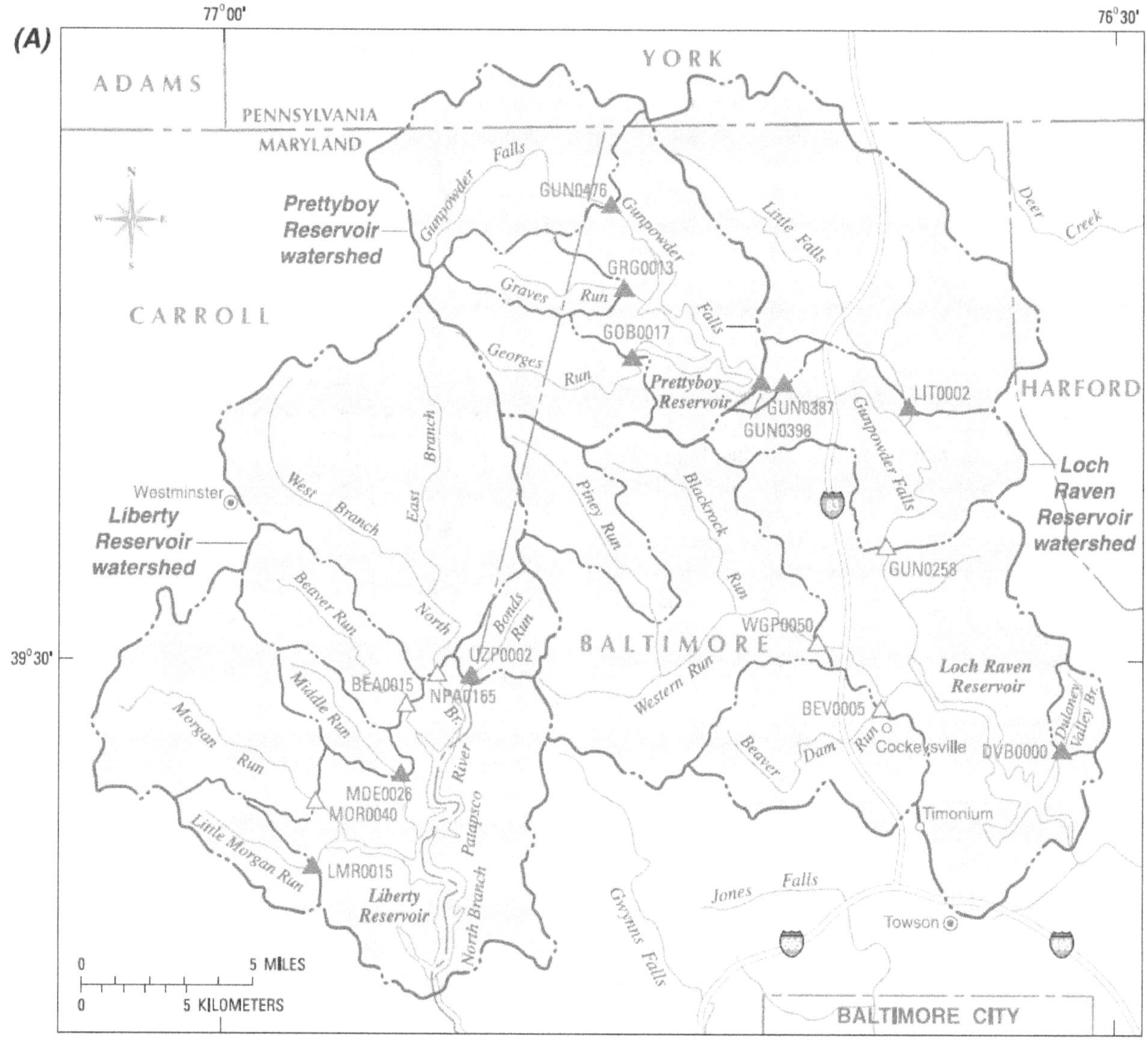

BASE FROM U.S. GEOLOGICAL SURVEY, 1:500,000

EXPLANATION

━━━ ••• ━━━ WATERSHED BOUNDARY

BEA0015 △ TRIBUTARY STATION FOR LONG-TERM
DRY-WEATHER AND STORMFLOW
MONITORING, IDENTIFIER, AND LOCATION

LIBERTY RESERVOIR
 Beaver Run at Hughes Road (BEA0015)
 Morgan Run at London Bridge Road (MOR0040)
 North Branch Patapsco River at Route 91 (NPA0165)

LOCH RAVEN RESERVOIR
 Beaver Dam Run at Beaver Run Lane (BEV0005)
 Western Run at Western Run Road (WGP0050)
 Gunpowder Falls at Glencoe Road (GUN0258)

━━━ •• ━━━ SUBBASIN BOUNDARY

LMR0015 ▲ TRIBUTARY STATION FOR LONG-TERM
DRY-WEATHER MONITORING, IDENTIFIER,
AND LOCATION

LIBERTY RESERVOIR
 Little Morgan Run at Bartholow Road (LMR0015)
 Middle Run at Louisville Road (MDE0026)
 Bonds Run at Hollingworth Road (UZP0002)

LOCH RAVEN RESERVOIR
 Dulaney Valley Branch at Loch Raven Drive (DVB0000)
 Gunpowder Falls at Falls Road (GUN0387)
 Gunpowder Falls below Prettyboy Dam (GUN0398)
 Little Falls at Blue Mount Road (LIT0002)

PRETTYBOY RESERVOIR
 Georges Run at Georges Creek Road (GOB0017)
 Graves Run at Gunpowder Road (GRG0013)
 Gunpowder Falls at Gunpowder Road (GUN0476)

Figure 3.   *(A)* Tributary and *(B)* reservoir monitoring stations in the Baltimore reservoir watersheds (modified from Baltimore City Department of Public Works, 2000 and Winfield and others, 2006).

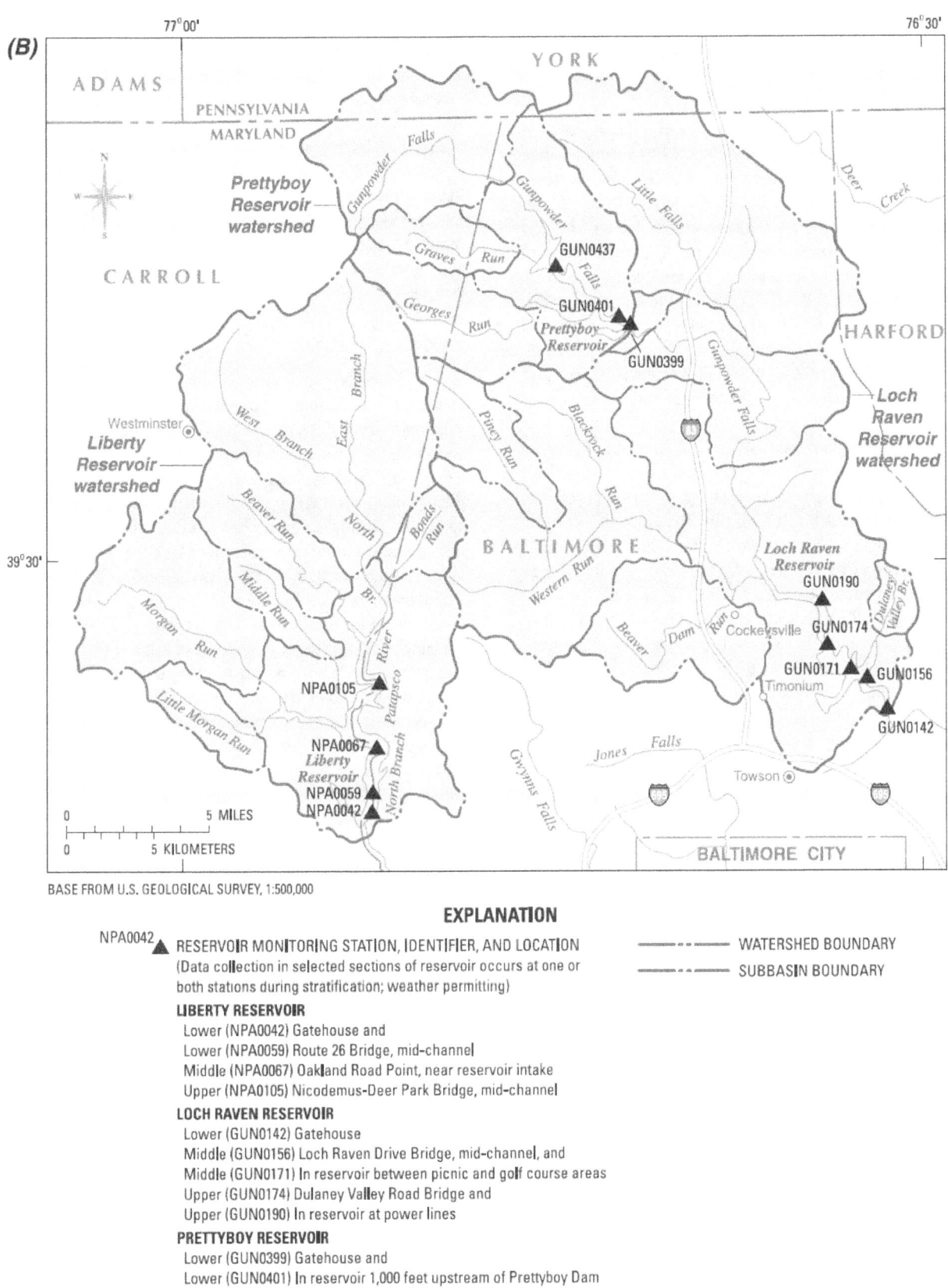

BASE FROM U.S. GEOLOGICAL SURVEY, 1:500,000

EXPLANATION

NPA0042 ▲ RESERVOIR MONITORING STATION, IDENTIFIER, AND LOCATION
(Data collection in selected sections of reservoir occurs at one or
both stations during stratification; weather permitting)

- - - - - - WATERSHED BOUNDARY
— - — - — SUBBASIN BOUNDARY

LIBERTY RESERVOIR
Lower (NPA0042) Gatehouse and
Lower (NPA0059) Route 26 Bridge, mid-channel
Middle (NPA0067) Oakland Road Point, near reservoir intake
Upper (NPA0105) Nicodemus-Deer Park Bridge, mid-channel

LOCH RAVEN RESERVOIR
Lower (GUN0142) Gatehouse
Middle (GUN0156) Loch Raven Drive Bridge, mid-channel, and
Middle (GUN0171) In reservoir between picnic and golf course areas
Upper (GUN0174) Dulaney Valley Road Bridge and
Upper (GUN0190) In reservoir at power lines

PRETTYBOY RESERVOIR
Lower (GUN0399) Gatehouse and
Lower (GUN0401) In reservoir 1,000 feet upstream of Prettyboy Dam
Middle (GUN0437) Beckleysville Road Bridge

Figure 3. *(A)* Tributary and *(B)* reservoir monitoring stations in the Baltimore reservoir watersheds (modified from Baltimore City Department of Public Works, 2000 and Winfield and others, 2006).—Continued

e) The suitability and adequacy of the areal extent of existing monitoring networks, and the need to sample additional areas in Carroll County or selected watersheds.

In 2007, the RTG prepared an interim report that described the status of selected 2005 RWMA Action Strategy commitments in relation to monitoring (Baltimore Reservoir Technical Group, 2007). This interim report indicated that the DPW will continue to conduct comprehensive water-quality monitoring in the three reservoirs and in selected major tributaries. As part of this monitoring, key pollutants of concern will be measured, and estimated annual loadings of sediment and total phosphorus will be calculated. The DPW also will continue to collect water samples, and modify its cooperative agreement with the USGS to include the development of annual load estimates for sediment and nutrients to Loch Raven and Liberty Reservoirs.

As of 2007, the design of the long-term monitoring network for water quality in the Baltimore reservoir system includes parts of each reservoir and selected watershed tributaries supplying each reservoir (fig. 3). Although monitoring conducted within each reservoir-watershed system generally is similar (**Appendix B**), there also are some notable differences in the extent and type of monitoring conducted.

In part, differences in monitoring reflect reservoir use. For the reservoirs used to directly provide drinking-water supplies, Loch Raven and Liberty, monitoring in three of the major tributaries for each reservoir includes storm- and dry-weather flows, and, within the reservoirs, includes stations near the water-supply intakes, and analysis of additional (raw and treated) water parameters. Monitoring in the tributaries in Prettyboy Reservoir, which re-supplies Loch Raven Reservoir, is limited solely to dry-weather flows. Also, no monitoring data are collected in the upper part of Prettyboy Reservoir. Because releases from Prettyboy Reservoir also are used to support recreational use (fish habitat) downstream, monitoring is conducted on the water released from Prettyboy Reservoir.

In part, differences in monitoring among the reservoirs reflect differences in the need and ability to access areas within each reservoir to characterize water quality. During periods of stratification (generally mid-spring through early fall), and weather permitting, sampling in each reservoir is conducted more frequently (twice rather than once per month), and, weather permitting, by boat at sites within selected areas of the reservoir, or from bridges and intake gatehouses. During periods of lake turnover, roughly late fall through early spring, and given the weather limits access to some areas within the reservoirs, sampling frequency and the breadth of constituents covered is reduced, and monitoring to address Federal requirements and State programs is mostly done from bridges and gatehouses.

Liberty Watershed and Reservoir

Long-term monitoring in Liberty Reservoir can be described in relation to selected watershed tributaries, the reservoir, and the Ashburton treatment facility. Monitoring at all three locations has been conducted since the early 1980s. All laboratory analyses of tributary, reservoir, and treatment facility samples traditionally have been conducted in the Ashburton treatment facility laboratory, and in accordance with their laboratory protocols (**Appendix C**, Quality-Assurance Plans, Ashburton treatment facility laboratory). From approximately 2005–08, renovations of this facility required analyses be performed by either a contract laboratory (chiefly tributary samples), or at the Loch Raven Reservoir Montebello treatment facility laboratory (chiefly reservoir samples).

Watershed Monitoring

City (Reservoir Natural Resources Section) staff conduct routine (monthly) dry-weather-flow sampling at six tributary sites (fig. 3), and, when discharging, at two NPDES sites within the watershed (**Appendix B**, table B1, Liberty Reservoir, Tributary and NPDES sites). In-field measurements taken at a single point in-stream with a multi-probe sonde include air and water temperature, pH, DO concentration (percent saturation is calculated), and specific conductance (**Appendix B**, table B2, Liberty Reservoir, Watershed Tributary Sites). Grab samples (at a single point in stream) are obtained for turbidity, solids, nutrients, alkalinity, and chlorides in low-density polyethylene bottles. If the site is co-located with a USGS streamgage, the stage height is recorded. The current-day and previous-day weather also are recorded.

Storm sampling is conducted by the City (Water-Quality Management Section) staff at three tributary sites for selected nutrients and solids (fig. 3 and **Appendix B**, table B2, Liberty Reservoir, Watershed Tributary Sites, parameters with E code). Although all three sites can be sampled for the same storm event, limitations in staff, equipment, and sampling logistics preclude sampling: (a) more than a few storms per site per year, (b) very large events, such as widespread flooding from a nor'easter or tropical cyclone, and (c) simultaneously for storms at these tributary sites and those tributary sites used for storm sampling in the Loch Raven Reservoir watershed. Sample collection also generally is limited to selected samples for solids and nutrients. Samples generally are obtained on an hourly basis with an ice-cooled automated sampler, which is temporarily installed before the anticipated storm event, and either activated after a pre-set elapsed time, or by a stream-stage activator. Storm sampling occasionally is conducted manually.

Reservoir Monitoring

City (Reservoir Natural Resources Section) staff conduct routine sampling in the lower, middle, and upper parts of the reservoir (fig. 3 and **Appendix B**, table B1, Liberty Reservoir, Reservoir sites) at monthly (winter) to bimonthly (spring-summer-fall) intervals for selected constituents (**Appendix B**, table B2, Liberty Reservoir, Reservoir sites). In the lower reservoir, sampling routinely is conducted throughout the year at Reservoir site NPA0042, and additionally at Reservoir site NPA0059 by boat and weather permitting, which generally occurs during the spring-summer-fall period (April through November or December). Sampling in the middle part of the reservoir only is conducted at Reservoir Site NPA0067, and by boat, weather permitting. Sampling in the upper reservoir always is conducted at Reservoir Site NPA0105. Data from paired stations in the lower parts of this reservoir generally are combined for interpretive analysis.

In-field data collection at each Liberty Reservoir site includes multi-probe sonde readings of water temperature, pH, DO concentration (percent saturation is calculated), and specific conductance. These data are obtained during profile sampling at approximately 5-ft intervals from just below the lake surface to a depth of 60 ft, and then at 10-ft intervals until the sensor strikes the lake bottom. (Bottom readings are not recorded.)

Algal and water-quality samples are collected for analysis at discrete depths at each reservoir monitoring site using a Kemmerer-style sampler. The samples for chl-*a* are collected at every site at 10-ft intervals starting from the surface to a depth of 50 ft. Samples for other chemical and additional analyses are obtained at each site as follows:

   a) Site NPA0042: Surface, 10-ft, an elevation of 365 ft (to correspond with the 55-ft deep intakes), and an elevation of 320 ft (to correspond with the 100-ft deep intakes);

   b) Site NPA0059 and Site NPA0067: Surface, and 10-, 20-, 40-, and 80-ft depths; and

   c) Site NPA0105: Surface, and 10-, 20-, and 40-ft depths.

Ashburton Treatment Facility Monitoring

Most of the data obtained by this City facility are used to enable water purveyors to assess recent and current (daily) water-quality conditions in order to help select intake depth(s) for raw-water supplies and determine the potential for nuisance problems in the treatment of this water or the finished water. Data used in this assessment include those routinely obtained by this facility (**Appendix B**, table B2, Liberty, Ashburton, TF, Raw) from samples delivered from gatehouse intakes. In addition, data used can include the most recent (monthly or bimonthly) and available reservoir water-quality data, mainly from the lower reservoir sampling locations (fig. 3 and **Appendix B**, table B1, Liberty Reservoir, Site NPA0042 and Site NPA0059). Sampling of treated water also is conducted by the Ashburton facility to assess residual chlorine and (or) the suitability of treated water for consumption (**Appendix B**, table B1, Liberty, Ashburton, TF, Treated). Most data obtained at this facility chiefly are used by water purveyors, but also ultimately can provide the RWMA partners with the most direct link between the quality of water in the lower Liberty Reservoir and the quality of treated or potable water.

Loch Raven Watershed and Reservoir

Long-term monitoring in Loch Raven Reservoir can be described in relation to selected watershed tributaries, the reservoir, and the Montebello treatment facility. Monitoring at all three locations has been conducted since the early 1980s. Traditionally, Loch Raven Reservoir dry-weather-flow samples have been analyzed by the Montebello treatment facility laboratory, in accordance with their laboratory protocols (**Appendix C**, Quality Assurance Plans, Montebello treatment facility laboratory). While the Montebello facility underwent renovations from 2005–07, samples were analyzed by either the Ashburton treatment-facility laboratory or a private contractor. Stormflow samples always have been analyzed by the Ashburton treatment-facility laboratory.

Watershed Monitoring

City staff (Reservoir Natural Resources Section) conduct routine (monthly) dry-weather-flow sampling at seven tributary sites (fig. 3), and, when discharging, at one irrigation pond and one NPDES site, within the watershed (**Appendix B**, table B1, Loch Raven Reservoir, Watershed, Tributary, Irrigation pond, and NPDES site). In-field measurements at a single point in-stream with a multi-probe sonde include air and water temperature, pH, DO concentration (percent saturation is calculated), and specific conductance (**Appendix B**, table B3, Loch Raven Reservoir, Watershed Tributary sites). Grab samples (at a single point in stream) are obtained for turbidity, solids, nutrients, alkalinity, and chlorides in low-density polyethylene bottles. If the site is co-located with a USGS streamgage, the stage height is recorded. Current-day and previous-day weather also are recorded.

Storm sampling is conducted by the City (Water-Quality Management Section) at selected (three) tributary sites for selected nutrients and solids (fig. 3 and **Appendix B**, table B1, Loch Raven Reservoir, Watershed Tributaries, parameters with E code). Although all three sites can be sampled for the same storm event, limitations in staff, equipment, and sampling logistics preclude sampling: (a) more than a few storms per site per year, (b) very large events, such as widespread flooding from a nor'easter or tropical cyclone, and (c) simultaneously for storms at these tributary sites and those tributary sites used for storm sampling in the Liberty Reservoir watershed. Samples are obtained in the same manner described above for storm sampling at selected watershed tributaries in Liberty Reservoir watershed.

Reservoir Monitoring

City staff (Reservoir Natural Resources Section) conduct routine sampling in the lower, middle, and upper parts of the reservoir (fig. 3 and **Appendix B**, table B1, Loch Raven Reservoir, Reservoir sites) at monthly (winter) to bimonthly (spring-summer-fall) intervals for selected constituents (**Appendix B**, table B3, Loch Raven Reservoir, Reservoir sites). In the lower reservoir, sampling routinely is conducted throughout the year at Reservoir Site GUN0142. In the middle and upper parts of the reservoir, sampling is seasonally divided between paired sites (**Appendix B**, table B3, Loch Raven Reservoir, Reservoir Sites: Middle—Site GUN0156 (winter, from bridge) and Site GUN0171 (spring-summer-fall by boat), and Upper—Site GUN0174 (winter, from bridge) and Site GUN0190 (spring-summer-fall, by boat). Data from paired stations, in either the upper or middle part of the reservoir, are combined for interpretive analysis of that part of the reservoir.

In-field data collection at each Loch Raven Reservoir site includes multi-probe sonde readings of water temperature, pH, DO concentration (percent saturation is calculated), and specific conductance. Data are obtained at intervals similar to those described above for Liberty Reservoir.

Algal and water-quality samples for analyses are collected at discrete depths at each reservoir site using a Kemmerer-style sampler. Samples for chl-*a* are collected at every site at 10-ft intervals starting from the surface to a depth of 50 ft. Samples for chemical and other analyses are obtained at each site at the surface and every 10 ft up to 60 ft.

Montebello Treatment Facility Monitoring

Most of the data obtained by this City facility are used to enable water purveyors to assess recent and current (daily) water-quality conditions in order to help select intake depth(s) for raw-water supplies and determine the potential for nuisance problems in the treatment of this water or the finished water. Data used in this assessment include those routinely obtained by this facility (**Appendix B**, table B3, Loch Raven Reservoir, Montebello, TF, Raw) from samples delivered from gatehouse intakes. This assessment also can include the most recent (monthly or bimonthly) reservoir water-quality data, chiefly from the lower reservoir sampling location (fig. 3 and **Appendix B**, table B1, Loch Raven Reservoir, Reservoir site GUN0142). Sampling of treated water also is conducted to assess residual chlorination, treatment results, and suitability of treated water for consumption (**Appendix B**, table B3, Loch Raven, Montebello, TF, Treated). As is the case for the Ashburton facility and Liberty Reservoir, data obtained at the Montebello facility is used primarily by water purveyors, but also could provide the RWMA partners with the most direct link between the quality of water in the lower Loch Raven Reservoir and the quality of treated or potable water.

Prettyboy Watershed and Reservoir

Long-term monitoring in Prettyboy Reservoir can be described in relation to selected watershed tributaries and the reservoir. Monitoring in the watershed and reservoir has been conducted since the early 1980s. Water-quality samples from Prettyboy Reservoir traditionally have been analyzed by the Montebello treatment facility laboratory, in accordance with their laboratory protocols (**Appendix C**, Laboratory Quality Assurance, Montebello treatment facility laboratory). While this facility undergoes renovation, samples are analyzed by the Ashburton treatment facility laboratory or a contract

laboratory. Tributary samples from the Prettyboy Reservoir watershed have traditionally been analyzed by the Ashburton treatment facility laboratory, except during its renovation from approximately 2005–08, when samples were analyzed by the Montebello treatment facility laboratory or a contract laboratory.

Watershed Monitoring

City staff (Reservoir Natural Resources Section) routinely conduct only (monthly) dry-weather-flow sampling at three tributary sites (fig. 3), and, when discharging, at one NPDES site within the watershed (**Appendix B**, table B1, Prettyboy Reservoir, Tributary and NPDES Sites). In-field measurements are made at a single point in stream with a multi-probe sonde. Measurements include air and water temperature, pH, DO concentration (percent saturation is calculated), and specific conductance (**Appendix B**, table B4, Prettyboy, Watershed Tributary sites). Grab samples (at a single point in stream) are obtained for turbidity, solids, nutrients, alkalinity, and chlorides in low-density polyethylene bottles. If the site is co-located with a USGS streamgage, the stage height is recorded. Current-day and previous-day weather also are recorded.

Reservoir Monitoring

City staff (Reservoir Natural Resources Section) conduct routine sampling in the lower and middle parts of the reservoir (fig. 3 and **Appendix B**, table B1, Prettyboy Reservoir, Reservoir, Tributary, and NPDES sites) at monthly intervals for selected constituents (**Appendix B**, table B4, Prettyboy Reservoir sites). Sampling in the lower reservoir is conducted throughout the year at Reservoir site GUN0399, and additionally at Reservoir site GUN0401 by boat and weather permitting—generally during the spring-summer-fall period (April through November or December). Sampling in the middle part of the reservoir is conducted throughout the year at Reservoir site GUN0437. Data from paired stations in the lower part of the reservoir generally are combined for interpretive analysis.

In-field data collection at each Prettyboy Reservoir site includes multi-probe sonde readings of water temperature, pH, DO concentration (percent saturation is calculated), and specific conductance. Data are obtained at intervals similar to those described above for Liberty Reservoir.

Algal and water-quality samples for analyses are collected at discrete depths at each reservoir site using a Kemmerer-style sampler. Samples for chl-*a* are collected at every site at 10-ft intervals starting from the surface to a depth of 50 ft. Samples for chemical and other analyses are obtained at each site as follows:

a) GUN0399 and GUN0401: Surface, and 10-, 20-, 40,- and 80-ft depths below the surface; and

b) GUN0437: Surface, and 10-, 20-, 40- and 60-ft depths below the surface.

Review and Evaluation

The effectiveness of the monitoring program for the Baltimore Reservoir System was assessed in part through a review of the monitoring database and data as of 2007. The effectiveness of the monitoring program was partially assessed on the basis of the most recent (2006) attempt at broad-based modeling to describe and relate water-quality conditions in the Loch Raven and Prettyboy Reservoir watershed tributaries to water-quality and biotic conditions in these reservoirs. The effectiveness of the monitoring program also was assessed on the basis of the use of monitoring data and information by RWMA partners, the RTG, and others to describe spatial and temporal variations in water-quality parameters associated with each of the long-term and emerging water-quality concerns.

Quality of the Monitoring Database and the Data Collected

A key measure of the value of a monitoring program that needs to provide data over the long term for analysis and decision-making is the quality of the database and the data within that database. Often a critical determining factor as to whether both of the above are adequate is a quality-assurance plan and program (QAPP) that is sufficiently comprehensive and actively followed to ensure the following:

a) A readily available complete and detailed description of the ongoing field and laboratory procedures and methods of data and sample collection for the monitoring program.

b) Definitions of the long-term data-quality requirements for the monitoring program, which include identification of the accuracy of the data (reporting level, precision, and bias) required or suitable for each constituent obtained by the monitoring program, and in particular, for those constituents that are used by the RTG as decision criteria (for example, to address SDWA or CWA 303d regulatory standards).

c) Verification measurements of data quality from quality-assurance and control (QAC) samples, and the manner in which the QAC data routinely are obtained and digitally stored with the corresponding water-quality data.

d) Description and implementation of the procedures for the routine (for example, annual or biannual) analysis, review, verification, and evaluation of data quality on the basis of the QAC data, and the documentation of findings, to determine if data requirements are being met.

e) Description and implementation of procedures for the systematic modification of field or laboratory methods, and the digital documentation thereof, given that data-

quality requirements, or methods or personnel used to obtain data under a long-term monitoring program invariably change.

The importance of the above elements in a QAPP cannot be underestimated. Long-term assessments of state, as well as trends, loads, and other interpretive measures of the physical, water-quality, and biotic conditions in the watershed tributaries and reservoirs related to RTG monitoring objectives are derived from analysis of long-term monitoring data. In turn, monitoring results guide RWMA action strategies and management decisions to meet RWMA goals.

In the absence of an adequate QAPP, it can be difficult to assess simple changes in state. For example, historically the RWMA partners and RTG had to rely on previous sediment bathymetric surveys to assess changes in reservoir storage capacity. Although several such surveys were conducted on the reservoirs, they were not adequately documented and archived in a manner that could be combined with recent RWMA surveys to determine periodic changes in reservoir storage capacities (Ortt and others, 2000). In addition, the lack of an adequate QAPP can result in considerable time and expense to address limitations in the quality of data if such limitations first become apparent during interpretative analyses.

In relation to the database and data obtained by the Baltimore Reservoir System long-term monitoring program, concerns such as those described above can be raised. The RTG has relied on biotic and chemical water-quality measurements and samples from the reservoirs and watersheds that are mainly collected by the City DPW Reservoir Natural Resources or Water-Quality Management Sections. The analyses of samples have been performed by the City at two different laboratories in the Ashburton or Montebello treatment facilities. Other contract laboratories also have been employed by the City for HAAs, suspended organic carbon, herbicides, metals, and cryptosporidium. Recent renovations involving both City water-treatment facilities (from 2005–09) resulted in first one and then the other treatment facility taking on added analyses, as well as an increased use of contract laboratories. With so many different groups involved in the collection and analysis of samples, the quality of the data being obtained by the monitoring program will likely come into question.

To determine the quality of database and data being used to address water-quality concerns, information was obtained from the City on the reservoir-monitoring locations and methods. A copy of the long-term monitoring database was obtained from the Baltimore City DPW, as were copies of the 2007 Montebello and Ashburton treatment-facility laboratory QAC plans (**Appendix C**). Reviews of the database and facility QAC plans were conducted in relation to each of the five elements of the comprehensive QAPP described above.

With respect to QAPP item a), and as of 2007, no single QAPP document was readily available from the RTG or City (DPW) that described in detail data- and sample-collection or sample-analyses procedures or protocols that included the monitoring locations, the types of data and samples collected, the frequency of their collection, and the specific methods, procedures, and equipment used in the field to obtain data and samples, or used by the laboratories to analyze samples. Thus, to aid the City, and this review, a fairly comprehensive description of the monitoring program (see **Current Perspective**, *this report*), and monitoring locations and data being collected by the City (**Appendix B**), were prepared with the assistance of City staff of the Water-Quality Management Section, DPW and brief descriptions found in historical reports. More detail is needed, however, on the specific procedures, methods, and equipment used to obtain samples in the reservoirs and watershed tributaries, and on the laboratories and methods used to analyze samples.

In relation to QAPP item b), and as of 2007, no description of specific data-quality objectives needed to meet the interpretive needs of the RTG or to guide agencies contracted to perform interpretive analyses for the RTG was found. Therefore, there are no clear data-quality requirements specified for the long-term monitoring data to guide the City DPW in the collection of data and samples or the City Ashburton and Montebello laboratories in the analysis of samples. That the lack of this information has occasionally affected the ability to analyze and interpret data is reflected in several studies that have noted selected data (historical or recent) were of unknown and questionable quality, and therefore, likely warranted caution in their use (Amatayakul, Defries, and others, 1978; Walker, 1988; KCI Technologies, Inc., 2004; Interstate Commission on the Potomac River Basin, 2006). Whereas some recommendations by these investigators appear to have been adopted, others have not been fully addressed. For example, clearly articulated data-quality requirements still do not exist, which could aid the RTG and City in the reduction, if not elimination, of these types of data-quality concerns, regardless of who ultimately provides the required data.

In relation to QAPP item c), a review of the 2007 QAC plans for both City laboratories (**Appendix C**) indicated that these plans are fundamentally different from what historically has been shown to be needed in a QAC plan for this long-term monitoring program. Both City laboratory plans focus chiefly on the performance of day-to-day analytical operations designed to assess the quality of water near or at the reservoir intakes for pending treatment, or the suitability of treated water for temporary storage, or immediate distribution and consumption. Data related to hydrodynamic conditions, such as reservoir water levels, are generally collected to meet these short-term needs, but not stored electronically. Water-quality samples are analyzed with Federal or State-approved procedures. However, these procedures are designed primarily for short-term needs to assess water-treatment requirements or compliance with Federal drinking-water standards, and are not generally well suited to the low concentrations of selected reservoir and watershed-tributary water-quality parameters—for example, total phosphorus. Analytical method changes also do not require extensive comparative testing between the new and old methods or long-term documentation of method changes and differences. For a long-term monitoring program, if such steps are not taken, the changes in data due

to method changes, and the lack of long-term documentation of such changes, can limit interpretations designed to assess changes in water-quality states or establish trends over decadal time periods (for example, see **Eutrophication—Nutrients: Phosphorus Transport and Reservoir Recycling**, *this report*). Also, archival of City laboratory QAC data is minimal if the analytical methods perform adequately that day; and, regardless of performance, these records are not required to be archived beyond a 5-year period. Hence, the City treatment-facility laboratory QAC management plans cannot be expected nor used to meet the QAC requirements of a long-term monitoring program.

Also in relation to QAPP item c), other QAC issues arose from the review of the long-term monitoring program database. Most notable are differences in the number of total (significant and insignificant) figures reported for a given analytical constituent, and the inability to appropriately "remark" certain types of analytical data. For example, concentration values for total phosphorus (TP), a constituent of considerable concern to the RWMA partners, are not reported to the same number of total figures within or among the three reservoirs, despite both City laboratories using the same analytical procedures and methods for sample analysis. Values appear rounded to fewer total figures for some samples than others. For other constituents, for example, DO, the number of significant figures is used to identify the analytical method used to obtain the concentration of DO, but does not truly reflect the precision of the measurement. Fundamentally, data for a given constituent technically should be systematically and independently remarked, and reported to the number of figures warranted by the precision of its measurement, as determined from QAC samples obtained during data collection and during analysis of samples. For example, the U.S. Geological Survey National Field Manual (variously dated) provides guidance on how quality-control samples and data for common water-quality measurements can be obtained, reported, and remarked, assuming operation of the in-field sampling or measurement equipment provides data that fall within the manufacturer-designated levels of precision.

The ability to adequately remark data for long-term storage, or, more importantly, enable knowledgeable retrieval of that data, is critical. For example, measurements that fall below the reporting level of an analytical method routinely are remarked as less than ("<"). The City DPW, however, adopted a procedure early on to enter such data as one-half the value of the analytical method reporting level. Over time there have been changes to analytical methods with lower reporting levels. Thus, it is no longer possible to identify all historical "less than" values. This has led to interpretive studies that identified such outlier data as highly unusual and unexplainable, and as jeopardizing the ability to interpret the entire dataset (Walker, 1998; KCI Technologies, Inc., 2004; Interstate Commission on the Potomac River Basin, 2006).

In relation to item QAPP item d), as of 2007 no indication was found during this review that systematic summaries, evaluations, and reviews of the quality of the data being collected are routinely (for example annually or bi-annually) performed and presented by the City as part of the monitoring program. A direct consequence of the above is that possible limitations in data quality can go undetected until long after data collection is complete, and only arise during subsequent independent analyses and interpretation of the data. In the absence of routine data-quality reviews, and because of the elapsed time involved, limitations in data discovered in this manner are unlikely to ever be clearly resolved (for example, see **Eutrophication—Nutrients: Phosphorus Transport and Reservoir Recycling,** and **Sedimentation—Sediment Transport**, *this report*).

In relation to item e), the responsibilities for the collection of data and samples in the field, as well as the analyses of the collected samples, have been conducted by different City staff or, for selected analyses, been contracted out to private laboratories over the course of long-term monitoring (early 1980s to present). However, there does not appear to be a concise description and manner of implementation for the systematic transfer of field or laboratory analysis and methods from one party to another, and the evaluation and documentation thereof.

Collectively, the review of the database and data for the monitoring program for the Baltimore Reservoir System indicates that as of 2007, there was no single document that could describe how this monitoring program addressed each of five basic elements described above in the QAPP for long-term water-quality monitoring. Shortcomings in the Baltimore Reservoir System monitoring program appear in relation to each of the five elements. These shortcomings appear to have affected the quality of data collected, and (or) the ability of the RTG, or selected contractors, to effectively use that data to address selected RWMA water-quality concerns.

Modeling to Address Water-Quality Concerns

The effectiveness of the monitoring network to support predictive tools, mainly computer models, to describe watershed and reservoir processes, or define TMDLs and related monitoring requirements, is critical to understanding and addressing long-term and emerging water-quality concerns and regulatory concerns (see **Overview of Water-Quality Concerns**, *this report*). Thus, there have been several modeling investigations directed toward relating water-quality conditions in the reservoir watershed tributaries to water-quality and biotic conditions in the reservoirs. Most notable in this regard was a recent study by the Interstate Commission on the Potomac River Basin (2006), which developed a modeling framework for simulating hydrodynamics and water quality in the Prettyboy and Loch Raven Reservoirs. Building upon earlier work by MDE, the Interstate Commission on the Potomac River Basin (2006) coupled a simulation model of the watersheds draining into Prettyboy and Loch Raven Reservoirs (Hydrological Simulation Program-Fortran, HSPF) with a two-dimensional simulation model to simulate the hydrodynamics and water quality of the reservoirs (U.S. Army Corps

of Engineers CE-QUAL-W2). The primary purpose of the coupled models was to link nutrient loads, in particular, phosphorus loads, from the reservoir watersheds to algal biomass concentrations, represented by chl-*a* concentrations, in the reservoirs. A secondary purpose was to calibrate the relation between autochthonous and allochthonous organic matter and DO concentrations in the hypolimnetic layer.

On the basis of comparisons of simulated and historical and long-term monitoring data from 1992–97, the Interstate Commission on the Potomac River Basin (2006) found and noted a number of limitations in the availability or quality of the long-term monitoring data in relation to their capability to simulate key reservoir water-quality constituents of concern, as follows:

a) **Reservoir temperature**: This physical parameter was among the most influential (and measureable) factors that affect the density of water and stratification, thus inhibiting the turbulent mixing between the epilimnetic, metalimnetic, and hypolimnetic layers. Among the most influential factors in the simulation of reservoir temperature was the elevation of the outflows from the reservoirs (measureable). During low-flow conditions into the reservoir, and high demands for water supplies, the elevation of outflows can be determined chiefly by reservoir withdrawals (measureable). Simulations were complicated by the lack of readily available data on the volumes of daily withdrawals for drinking water, and, when the reservoirs were at capacity, the lack of data on the daily volume of water releases from the supply reservoirs (measureable).

b) **Reservoir water quality**: The primary purpose of the water-quality simulation was to calibrate the relation between chl-*a* in each reservoir and TP loads from each reservoir watershed. A secondary purpose was to calibrate the relation between autochthonous and allochthonous organic matter and DO in the hypolimnetic layer. Four stages were used to simulate and validate these relations, which involved reservoir concentrations for TP (measured as orthophosphate phosphorus, in a persulfate-digested, raw-water sample), chl-*a*, DO, and ammonia- and nitrate-nitrogen (all four parameters being generally measureable and routinely monitored):

1) *Total phosphorus*: Concentrations of TP were simulated for reservoir surface and bottom conditions from simulated watershed TP loads in combination with reservoir settling rates for suspended sediment. Although simulated tributary loads for TP were reasonably accurate, the lack of tributary data for stormflows, particularly in the Prettyboy Reservoir watershed, limited model calibration and validation. Simulated reservoir surface and bottom concentrations for TP did not consistently match the variability in observed TP concentrations in either

layer in either reservoir. In particular, monthly average surface TP concentrations were overestimated in 1993 and 1994, particularly for late fall storms in 1993, and notably underestimated in relation to major reservoir inflow events in 1996 and 1997. The inconsistencies were considered to be a function of the quality of the TP data and the quality of the model simulation. Regarding the latter, no data were available on dissolved organophosphate phosphorus (DOP, measured as orthophosphate phosphorus, in a filtered water sample) for model calibration, but DOP is the standard state variable for phosphorus in the reservoir water-quality model.

2) *Chlorophyll-a*: Concentrations of chl-*a* in the epilimnion were simulated for three seasons each year: (a) a mixed-species winter assemblage, (b) dominant spring algal taxa, and (c) dominant summer-fall algal taxa. The goal of the calibration was for the simulated chl-*a* concentration in each season to be at least as large as the observed chl-*a* concentration in that season (but not necessarily match by date and time). The resulting calibration was considered conservative. Ensuring that the simulated chl-*a* peaks are at least as large as observed peaks helps guarantee that the models can be used to calculate what TP loads are compatible with the reservoirs meeting water-quality standards. Growth rates and temperature coefficients were varied in the simulation by year and season, reflecting the variety of dominant algal species and the variety of factors that determined species succession. Although the goals of the simulation were met, a wide variety of different algal species can assume dominance in Loch Raven Reservoir, and most every spring and summer season is dominated by a different species. (Although it was not specified, the variability in algal dominance could indicate that variations in the amount of available phosphorus (and (or) nitrogen) dictate what species dominates and the duration of its occurrence.) Notable in the case of both reservoirs was a lack of data on winter and early spring algal counts, and, for Prettyboy Reservoir, on taxa identification.

3) *Dissolved oxygen*: Concentrations of DO were simulated for reservoir epilimnetic and hypolimnetic layers. Simulations were heavily dependent on the simulation of temperature, which determines density differences that inhibit oxygen transport through turbulent diffusion. Under stratification, simulated concentrations of DO in the water column were determined in relation to oxygen demands associated with the decomposition of labile dissolved organic matter (from autochthonous sources, algal detritus, and from allochthonous sources, tributary labile organic matter) and dissolved chemical biological oxygen demand. Nitrification of ammonia also

was considered. Under stratification, bottom DO concentrations were assumed to be determined by sediment oxygen demand, and represented by a temperature-dependent first-order decay process of a single type of organic matter. Simulations of surficial and bottom DO matched observed DO in relation to seasonal trends and average monthly concentrations. Simulations tended to underestimate surface DO concentrations in Loch Raven Reservoir. Overall, simulations of chemical or chemical-biological oxygen demands could not be verified due to the lack of actual measurements on these oxygen demands.

4) *Ammonia- and nitrate-nitrogen*: Concentrations of ammonia- and nitrate-nitrogen generally were assumed to occur in excess of aquatic needs, and available phosphorus was considered the limiting nutrient. In both Prettyboy and Loch Raven Reservoirs, approximately 93 percent of the surficial monitoring samples (10- and 20-ft depths) had nitrate-nitrogen to TP ratios of 10 or greater. The median nitrate-nitrogen to TP ratio in Loch Raven Reservoir was 38 and the median ratio in Prettyboy Reservoir was 47. The Interstate Commission on the Potomac River Basin (2006) concluded monitoring data overwhelmingly indicate that phosphorus is the limiting nutrient. Nevertheless, the simulation of nitrogen species was calibrated against the observed data. In general, model simulations had limited success in capturing the broad seasonal variability in either nitrate or ammonia nitrogen concentrations in either surface or bottom layers in either reservoir. Simulations also failed to capture the intra-seasonal variability concentrations of either nitrogen species in either surficial or bottom waters in either reservoir. The greatest deviations between simulated and observed values tended to occur in relation to extreme events—for example, in relation to storm events, or during very wet or dry years. A key limitation in the simulation of these nitrogen species was the lack of data on organic nitrogen.

The Interstate Commission on the Potomac River Basin (2006) noted that accurate simulations of water-quality conditions in Prettyboy and Loch Raven Reservoirs were limited by the lack of readily available data, or simply lack of data for selected hydrodynamic and biological parameters [see items 1) through 4) above]. Applying these limitations to modeling all three reservoir watersheds and reservoirs would indicate the following data are not obtained, obtained at too low a frequency, or obtained but not readily available:

a) Routine (daily) reservoir withdrawals (from Loch Raven and Liberty Reservoirs, not readily available), releases (from Loch Raven and Liberty Reservoirs, unavailable), and water levels from all three reservoirs (not readily available);

b) Routine measurements of reservoir water temperature (too low a frequency, requiring at least daily), which was found to be the most influential model parameter for the determination of reservoir stratification (epilimnion, metalimnion, and hypolimnion) or mixing;

c) Routine algal counts and taxa identification in winter and early spring [clearly a problem for Prettyboy Reservoir given evidence (chl-a) of blooms in late winter and early spring];

d) Routine measurements (unavailable) for DOP, total organic nitrogen, total organic carbon, and chemical and (or) biological oxygen demand;

e) Increased frequency of routine measurements for chl-a (daily) and DO (possibly diurnal), given the TMDL endpoint criteria (see **Regulatory Concerns**, *this report*).

The above measurements have been utilized in other model simulation studies developed with the U.S. Army Corps of Engineers CE-QUAL-W2 two-dimensional reservoir model to provide reasonably accurate simulations of water-quality and biotic conditions associated with seasonal algal blooms (Giorgino and Bales, 1997; Bales and Giorgino, 1998; Sarver and Steiner, 1998; Bales and others, 2001; Galloway and Green, 2004, 2006 a, b). All of these studies were able to simulate seasonal moderate to major algal blooms and their associated pre-, concomitant-, and post-bloom water-quality conditions in different lakes and reservoirs with reasonable accuracy and reliability. In all of the studies, simulations relied on the collection of reservoir data at similar to often higher frequencies (daily to monthly) than the frequency of data collection used in the long-term monitoring program for the Baltimore Reservoir System (generally monthly or bimonthly). In addition, these studies relied on measurements of DOP rather than just TP, as the former varied at times independently of the latter. It also was evident in several of these studies that DOP sources other than tributary inflows—for example, releases from nearby WWTPs and (or) reservoir bed sediments or previous phytoplankton blooms, likely contributed to phytoplankton production, and that reservoir DOP concentrations were not equal, or always directly proportional, to TP during the spring, summer, or fall, when phytoplankton blooms were most likely to occur.

Monitoring to Address Individual Water-Quality Concerns

The ability of the monitoring program for the Baltimore Reservoir System to provide data for the reservoir watershed tributaries to (a) describe the state, and temporal variations in state—such as seasonal, annual, and long-term trends, in water quality (b) describe the areal extent of pollutant sources and their possible relation to land use and cover in the watersheds, and (or) (c) relate water-quality conditions in the tributaries

to water-quality and (or) biotic conditions in the downstream reservoirs (fig. 3) were used to assess the effectiveness of this program to address each long-term and emerging RWMA concern. This assessment was conducted on the basis of a retrospective review of the 1980 through 2007 scientific and technical reports provided by the RTG partners or their contracted investigators.

Long-Term Water-Quality Concerns

Long-term water-quality concerns addressed in this review and evaluation and described by the 2005 and earlier RWMAs include eutrophic conditions in the reservoirs that result in major algal blooms, which are attributed to elevated nutrient (nitrogen and phosphorus) loads from tributaries upstream of the reservoirs. In turn, algal bloom die-offs and decomposition can promote deep-water anoxia in the reservoirs, and the subsequent release of iron and manganese from reservoir bed sediments, which in addition to algal residues, can interfere with the treatment of reservoir water and quality of treated water for drinking water, and thus are of concern. Sedimentation is an additional long-term concern. It reflects overland and tributary in-stream bed and bank erosion, which degrade tributary stream-water-quality and biotic conditions and adversely affect the designated recreational uses of streams. The resultant sediment loads also reduce the storage capacity of the water-supply reservoirs, which impairs their designated use for such supplies, and is of concern. Transported sediment also is the major source of metals, including iron, manganese, and mercury, and a source of nutrients, including phosphorus, in the reservoirs. Mobilization of these metals and phosphorus is a concern. Fecal coliform and other bacteria are another long-term concern. Although these bacteria have not been found at concentrations that exceed State standards for recreational activities in the reservoirs, they have been found at elevated concentrations in selected tributaries, and their occurrence is cause for concern given the designated recreational uses that involve water contact in the tributaries.

Eutrophication

Eutrophication is the process that enhances the production of algal and higher plants in response to enrichment by the plant nutrients nitrogen and phosphorus (Phillips, 2010). In assessing eutrophic conditions and their impacts on the Baltimore Reservoirs, the reservoirs always have been and remain subject to nutrient enrichment and eutrophication (see **Overview of Water-Quality Concerns**, *this report*). Therefore, in addressing eutrophication in the Baltimore Reservoirs, it is important to note that the reduction in the occurrence of eutrophic conditions that result in major algal blooms is the primary goal for the RWMA partners.

Summary (Baltimore City Department of Public Works, 1996, 2001; Baltimore Reservoir Technical Group, 2004) and other reports (Valcik, 1975; Winfield and Sakai, 2003; Interstate Commission on the Potomac River Basin, 2006) have utilized algal counts and taxa identification data from the long-term monitoring program (**Appendix B**) to describe phytoplankton production and bloom characteristics in the reservoirs. These reports all generally indicate that although algal counts appear to have declined (at least over the period from 1980–2001, and particularly through the 1990s), all three City reservoirs appear to remain subject to the effects of ongoing eutrophication. The use of the terms "apparent" or "appear to" here and throughout the remainder of this report indicates reported trends, seasonal patterns, or other relations where no statistical analyses were used to verify the trend, pattern, or relation.

Under what might be best termed typical mesotrophic conditions, which generally can be defined as average nutrient conditions, and average to above-average clarity, the aforementioned reports indicate phytoplankton production increases in the spring as daylight length increases and water temperature rises, and thermal stratification of each reservoir begins. Production generally builds to a maximum in the shallow reservoir waters—defined by the City, and for the purposes of this report, as waters up to 20–30 ft below the reservoir water surface—in the late summer to early fall, and then generally declines with reduced daylight length and temperature and reaches a minimum by late fall or early winter, coincident with lake turnover.

Long-term (1992–2004) data for Loch Raven and Prettyboy Reservoirs on routine algal counts and taxa identification from reservoir monitoring (**Appendix B**) were analyzed by the Interstate Commission on the Potomac River Basin (ICPRB) (2006). Their report indicated that for Loch Raven Reservoir, the relative abundance of taxa follows a typical seasonal spring-summer-fall succession. Golden-brown algae and diatoms generally dominate in the spring (February through April), and can continue their dominance until about mid-summer (approximately July). Green algae then briefly dominate until late summer to early fall (August through September), when, absent conditions for a major algal bloom, diatoms resume dominance. Throughout the summer, dinoflagellates can account for 15–25 percent of the algal count.

Also, according to the ICPRB report (2006), and except for the lack of occurrence of golden-brown algae, Prettyboy Reservoir exhibits a spring-summer-fall seasonal succession in the relative abundance of taxa similar to that described for Loch Raven Reservoir. In addition, although data on algal count and taxa identification only routinely were collected from these reservoirs from April or May through September, and sampling in Prettyboy Reservoir is only conducted monthly, the ICPRB noted that other monitoring data (chl-*a*) indicate major phytoplankton blooms (taxa and counts unknown) occur in Prettyboy Reservoir in late winter to early spring.

Long-term data on routine algal counts and taxa identification from reservoir monitoring for Liberty Reservoir (**Appendix B**) were analyzed by Winfield and Sakai (2003). Their results indicate that phytoplankton production in this

reservoir typically follows a bimodal (spring and late-summer-fall) seasonal succession. Surface-water inflows and complete mixing of the reservoir during the winter are considered the source of nutrients for a spring bloom (February through April or May), which typically is dominated by diatoms and green algae. As the reservoir stratifies, phosphorus in the shallow reservoir waters is reduced, and temperatures in shallow waters become warmer. These conditions are considered by Winfield and Sakai (2003) as unfavorable to continued phytoplankton in the spring bloom, which declines in production (generally May through June). Decomposition of the spring bloom, in combination with nutrient loads from surface-water inflows, is presumed to provide the nutrients (chiefly phosphorus) for the late summer bloom, dominated by blue-green and green algae, that begins mid-summer and continues in the late summer and fall (August through about October).

Winfield and Sakai (2003) also noted that the bimodal pattern in phytoplankton production is observed throughout Liberty Reservoir. In their report, however, algal counts and chl-*a* data indicate that the magnitude in productivity clearly follows a longitudinal gradient—greatest in the upper part, and smallest in the lower part, of the reservoir. By way of contrast, no studies were encountered during this review that addressed in detail whether or not marked spatial differences occur in the relative abundance (by counts) or type of taxa in either Loch Raven or Prettyboy Reservoirs.

Whereas the studies by ICPRB and Winfield and Sakai describe general seasonal phytoplankton production and succession characteristics for the three reservoirs, excessive algal-bloom production under eutrophic (elevated nutrient) conditions and its impact on reservoir water quality have only been described in detail for Liberty Reservoir. Under the right conditions, which Winfield and Sakai (2003) indicate relate to phosphorus availability, excessive blooms occur in Liberty Reservoir. Most often, these major blooms involve the typical taxa—green algae in mid-summer, and (or) blue-green algae in the late summer and early fall (Valcik, 1975; Winfield and Sakai, 2003; Baltimore Reservoir Technical Group, 2004; Interstate Commission on the Potomac River Basin, 2006). Less common major blooms have occurred in the fall, winter, or spring. For example, during the latter phase of the drought in the 1960s, and possibly related to excessive reservoir drawdown, major diatom blooms occurred in Liberty Reservoir in the spring of 1967 and spring of 1968; a major golden-brown algal bloom occurred in Loch Raven Reservoir in the spring of 1996.

Winfield and Sakai (2003) created a conceptual depth profile that illustrates temporal water-quality conditions related to excessive algal blooms in Liberty Reservoir (fig. 4). This illustration can serve as a model of this phenomenon for all three reservoirs, although the taxa, and actual timing, spatial extent, and magnitude of the changes in water quality that result from a major bloom can vary among the reservoirs.

In Liberty Reservoir, major blooms are most often distinguished by elevated algal counts and high concentrations of

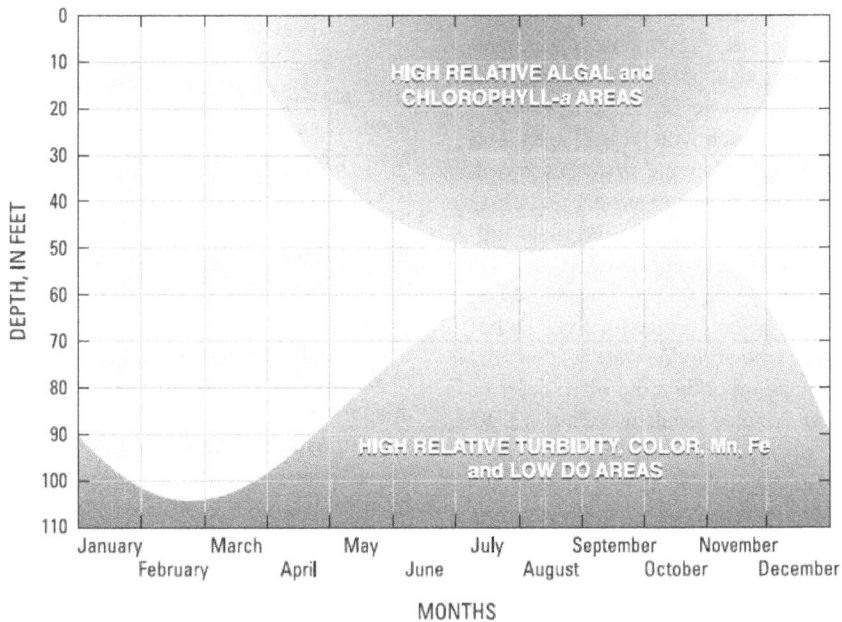

Figure 4. Composite generalized seasonal depth profile of algal-impaired water-quality conditions in Liberty Reservoir—from turbidity, algal-count, color, and concentration data for chlorophyll-*a*, dissolved oxygen (DO), manganese (Mn), and iron (Fe) (modified from Winfield and Sakai, 2003).

chl-*a* in shallow reservoir waters. Phytoplankton photosynthesis during these blooms can lead to (daytime) supersaturated DO concentrations and increases (spikes) in pH in the shallow reservoir waters. Algal respiration, and ultimately die-off and residue decomposition that follow a major bloom, appear to create a high biochemical oxygen demand, which Winfield and Sakai (2003) indicate leads to oxygen and pH sags (reduced concentrations of DO and lowered pH), and high color and turbidity, which generally are most pronounced in deep reservoir waters—here, and hereafter, in this report, defined as waters 30 ft or more below the reservoir water surface—in the late fall and early winter (fig. 4). With lake turnover in the late fall or early winter, only the deepest reservoir waters remain highly turbid into the late winter and early spring.

Winfield and Sakai (2003) also note that the decomposition of algal residues does appear to occasionally lead to anoxic conditions in surficial bottom sediments, which result in the release of manganese and iron presumably by sediment diagenesis. This release would account for the elevated concentrations of these two metals often observed in reservoir intake waters after excessive blooms die off, and DO declines at deep depths, as these metals are naturally abundant in origin, and found at elevated concentrations in all reservoir

sediments (Ortt and others, 2000). In addition, iron mobilization is thought to be accompanied by the release of phosphorus, which Winfield and Sakai (2003) also noted was observed for a number of years at elevated concentrations primarily in deep reservoir waters following major blooms. As in the case of manganese and iron, reservoir bottom sediments also contain abundant phosphorus, which largely is bound to iron (Ortt and others, 2000).

Ultimately, algal bloom die-off and decomposition that leads to hypoxic to anoxic conditions also is assumed by the Baltimore Reservoir Technical Group (2004) to produce physical and chemical conditions that adversely affect biota, including higher trophic sport-fish populations. In addition, the physical and aesthetic conditions associated with blooms and their decay is assumed to reduce recreational opportunities such as fishing and other water-contact activities, which are designated uses for all three reservoirs.

In addition to the impact on reservoir biota and recreational uses, the impact of excessive blooms on drinking-water supplies is also of concern. Although such blooms tend to occur in association with, they are not limited to, reservoir recovery after droughts and major drawdowns (Valcik, 1975; Winfield and Sakai, 2003; Baltimore Reservoir Technical

Group, 2004). Whether their occurrence is widespread, or simply near the water-supply intakes, eutrophic conditions (fig. 4) that result in excessive algal blooms can reduce the quality and quantity of reservoir water available for potable supplies. Among the taxa associated with these blooms, blue-green and green algae appear to create the broadest array of water-treatment problems and subsequent nuisance complaints regarding odor, color, and taste (Valcik, 1975; Winfield and Sakai, 2003; Baltimore Reservoir Technical Group, 2004).

Winfield and Sakai (2003) indicate reservoir withdrawals of water during or after such blooms preferentially are made from the intake(s) whose depth(s) provide water(s) that are the least affected by algal blooms or their die-off in order to reduce treatment costs and possible problems with color, odor, and (or) taste of treated water. Withdrawing such water in sufficient quantities, however, can be difficult, and occasionally not possible. Major algal blooms can occur from May through November, depending on the supply reservoir, which covers the period of highest annual daily demands—late spring through early fall. The range of depths at which relatively unimpaired supply waters are available can narrow markedly through time (fig 4.), however, as a major bloom progresses, dies off, and decomposes, and oxygen depletion related to decomposition results in elevated concentrations of metals at depth. Elevated concentrations of algal matter in any state at any depth, and elevated concentrations of iron and manganese at depth, both interfere in the treatment of water, and with treated water color, odor, and (or) taste.

Overall, the long-term monitoring network has provided the City and the RTG with the data to generally (a) describe seasonal phytoplankton production during lake stratification in each reservoir and either mesotrophic conditions or eutrophic conditions associated with major algal blooms, (b) relate broad seasonal patterns in the occurrence of phytoplankton (taxa and abundance) to their effect on other current or subsequent water-quality conditions (measured parameters) in each reservoir, and (c) identify bloom-related water-quality impairments to recreational uses in the reservoir under eutrophic conditions with major algal blooms. Under these conditions, and in the case of the water-supply reservoirs, the City also can identify the related water-quality impairments to supplies, and operate supply-reservoir intakes to reduce bloom effects on water-treatment costs and the quality of treated water.

Long-term monitoring data, or the analysis of these data, have not enabled the RTG to relate changes in phytoplankton taxa or abundance to water-quality conditions antecedent to their occurrence. Thus, water-quality conditions such as the concentrations of phosphorus and (or) nitrogen in the reservoir that lead to excessive algal blooms, or specific types of taxa blooms, are unknown. Long-term monitoring also has not adequately addressed, through algal taxa and counts, the blooms that occur in the winter and early spring in Prettyboy Reservoir. Nor could it be established via any received RTG report that monthly sampling for algal and taxa counts is sufficient to accurately characterize phytoplankton abundance or succession, particularly in Prettyboy Reservoir.

Nutrients

Eutrophic conditions that result in excessive algal blooms are considered by the RTG (2004) to be caused by excessive levels of nutrients entering a reservoir, and the amounts of nutrients (nitrogen and, in particular, phosphorus) transported from the Loch Raven (including Prettyboy) and Liberty Reservoir watersheds to their respective reservoirs are therefore a concern (fig. 2). It also should be noted that the occurrence of elevated concentrations of nutrients in streams have been cited as a possible contributing factor to excessive algal blooms in selected major tributaries (Valcik, 1975), which also could affect stream aquatic habitats and designated recreational uses, such as fishing.

To reduce nutrient concentrations in streams and loads to the reservoirs requires quantifying their transport within and from the reservoir watersheds, identifying their main sources within the watersheds, and reducing nutrient concentrations from these sources. Data from long-term monitoring in the watershed tributaries has been used to estimate nutrient loads, concentrations, and trends, and identify possible source areas on a subbasin scale.

Nitrogen Transport and Reservoir Cycling

Elevated nitrogen concentrations could play a role in phytoplankton production. Although Amatayakul and others (1978) concluded that phosphorus was the limiting nutrient in Baltimore reservoir waters, their greatest increase in assay algal production was in response to samples in which both concentrations of nitrogen and phosphorus were elevated (table 3). Also, increases in nitrate contributions from known point sources (WWTPs) to tributary streams were observed in the mid-1980s, just before major blooms occurred in Loch Raven and Liberty Reservoirs (Baltimore City Department of Public Works, 1996).

Before implementation of the long-term monitoring network, Amatayakul, Defries, and others (1978) provided and compared estimates for total nitrogen loads with earlier load estimates for Loch Raven Reservoir (table 4). They noted a lack of spatial and temporal coverage in the monitoring data; for example, nitrogen data only were available in subbasins representing 27 percent of the reservoir watershed area. Thus, their load estimates for the entire watershed were largely derived in the absence of actual measurements of nitrogen in most subbasins. In addition, they cited several known WWTPs whose effluents likely contributed to the nutrient load but also were unmeasured. Their monitoring recommendations for nutrients included year-round data collection at known major point-source discharges and in storm- and base flows on all major subbasins in the reservoir watershed.

The long-term monitoring network created in the early 1980s did include monitoring at known major point sources in the major tributaries for nitrogen but chiefly in forms available for biological uptake—ammonia, nitrate, and nitrite nitrogen, hereafter referred to as available forms of nitrogen in

Table 3. Bioassay results for nutrient-spiked samples of *Selanastrom capricornatum* (from Amatayakul, Defries, and others, 1978).

[mg/L, milligrams per liter]

Nutrient-spiked sample concentration (mg/L)	Maximum algal yield (mg/L, dry weight)
Control	1.0
Phosphorus 0.05	8.1
Phosphorus 0.05, Nitrogen 1.00	14.7
Nitrogen 1.00	1.2

Table 4. Estimated average annual loads of nitrogen and phosphorus to the Baltimore reservoirs, various periods of record between 1973–97.

[kg/yr, kilograms per year; ---, unavailable]

Constituent	Reservoir/ watershed	Estimated average annual load to the reservoir on basis of cited water years (kg/yr x 1,000)					
		1973–74[1]	1975–78[2]	1983–19--[3]	1983–90[4]	1988–91[5]	1992–97[6]
Nitrogen	Liberty	---	---	---	---	---	---
	Loch Raven	625	774, 744	---	---	3,450	717
	Prettyboy	---	---	---	---	---	287
Phosphorus	Liberty	---	---	31.2	44.6	---	---
	Loch Raven	11.1	14.7, 21.3	48.9	64.0	49.0	49.7
	Prettyboy	---	---	---	17.6	---	22.9

[1] Amatayakul, Defries, and others (1978), National Eutrophication Study, Report for Loch Raven Reservoir, Environmental Protection Agency, Region III, Working Paper No. 358, Corvallis, Oregon—likely few stormflows, load estimates for ungaged subbasins, with loads derived from normalized flow-concentration curves, and annual mean concentration.

[2] Amatayakul, Defries, and others (1978), includes some stormflows, load estimates for ungaged subbasins, with load estimates derived from flow-duration and concentration curves (first value) or annual mean flow and concentration (second value).

[3] Baltimore City Department of Public Works (1992), computed from statement that 1983–90 load estimates (see footnote 4, below) reflect a 30-percent increase for Loch Raven Reservoir and a 43-percent increase for Liberty Reservoir compared to prior published estimates. No information is available on how these prior estimates were computed, or publication source or date.

[4] Baltimore City Department of Public Works (1992), includes stormflows, load estimates for point and nonpoint sources, and assumes estimates from flow-duration and concentration curves.

[5] Maryland Department of the Environment (2003d), includes stormflows, load estimates for total phosphorus from point and nonpoint sources, including gaged and ungaged subbasins, for 10-year synthetic period under existing conditions generated from U.S. Environmental Protection Agency Storm Watershed Management Model, calibrated and validated with data from water years 1988–91.

[6] Interstate Commission on the Potomac River Basin (2006), load estimates from point and nonpoint sources for ungaged subbasins created with the Hydrological Simulation Program–FORTRAN (HSPF) model, and calibrated with data from water years 1992–97.

this report. As of 2007, no or few data have been collected on Kjeldahl or organic nitrogen.

The RTG and others have used data on the available forms of nitrogen and improved models to provide more accurate estimates of total annual loads of these nitrogen species to the reservoirs (table 4); these estimated loads generally are greater than the loads initially estimated by Amatayakul, Defries, and others (1978).

Monitoring data also have been used to describe and compare concentrations of available forms of nitrogen from point sources (NPDES includingWWTPs) among subbasins within a given reservoir watershed, or, on the basis of data from dry-weather-flow stations, as form-specific areal nitrogen loads. For example, upgrades to the WWTPs to reduce effluent nutrient loads were begun in the late 1970s and continued through the 1980s. Using data provided by the long-term monitoring network for NPDES sites, the City reported that initially there appeared to be no substantial reductions in total available nitrogen, but that there were shifts in the proportion of the forms of available nitrogen—from ammonia to nitrite and (or) nitrate nitrogen in WWTP effluents. By the mid-1990s, however, the monitoring data indicated a reduction in total available nitrogen loads at two WWTPs—Dutterers and Manchester—in Prettyboy and Loch Raven Reservoir watersheds, and an increase in load at a third WWTP—Hampstead (Baltimore City Department of Public Works, 1996, fig. 9.1). Monitoring in Liberty Reservoir watershed also indicated total available nitrogen loads eventually declined at its three WWTPs—Congoleum, Montrose, and Roy F. Weston (Baltimore City Department of Public Works, 1996, fig. 9.2). The documented reductions in available nitrogen from WWTP discharges demonstrated the effectiveness of the long-term monitoring program to track changes in known point-source loads for the available forms of this nutrient through the USEPA NPDES program.

Data obtained from the long-term monitoring network have been used to indicate apparent annual trends in ammonia, and apparent annual trends and seasonal variations in nitrate, nitrogen in the watershed tributaries. Decreases in annual median ammonia-nitrogen concentrations were observed in dry-weather flows at all tributary monitoring stations during the 1980s (fig. 5), which continued through the 1990s (Winfield and Sakai, 2003). The apparent declines in ammonia-nitrogen concentrations for most tributary stations in the Loch Raven Reservoir watershed eventually were verified to be statistically significant (Maryland Department of the Environment, 2003d, Kendall trend test, accounting for autocorrelation, p less than 0.05, data from 1981–95).

During the 1980s through the mid-1990s, concentrations of nitrate nitrogen appeared to increase at most tributary monitoring stations, and more so in the Liberty Reservoir watershed than in either the Loch Raven or Prettyboy Reservoir watersheds, but by 2000, they had appeared to level off (fig. 6). The apparent increasing trends in nitrate-nitrogen concentrations for the tributary stations in the Loch Raven Reservoir watershed eventually were statistically verified by MDE (Maryland Department of the Environment, 2003d).

Dry-weather flow monitoring data have been used to indicate seasonal patterns in tributary nitrate-nitrogen concentrations. Except for those tributaries overly influenced by WWTP effluent, or Prettyboy Reservoir outflow (for example, at Gunpowder River at Falls Road), mean-monthly nitrate-nitrogen concentrations for dry-weather flows in watershed tributaries cycle from winter highs to summer lows—a pattern attributed to greater riparian or in-stream biotic uptake or abiotic sequestration of nitrate during summer compared to winter months (Baltimore City Department of Public Works, 1996, fig. 9.11).

Monitoring data for dry-weather flows have been used to relate differences in nitrate-nitrogen concentrations among streams to differences in land cover. The City illustrated that long-term median nitrate-nitrogen concentrations in a stream

appeared to decrease as a function of the amount of forested land in the subbasin draining to that stream (Baltimore City Department of Public Works, 2001, fig. 44). No statistical analyses were conducted, however, to determine the strength of this relation or its level of statistical significance.

Given the aforementioned apparent or confirmed annual trends observed in ammonia- and nitrate-nitrogen concentrations in the reservoir watershed tributaries, the question arose as to whether similar trends could be discerned in the reservoirs. Overall, City DPW summary and other reports (Baltimore City Department of Public Works, 1996, 2000, 2001; Winfield and Sakai, 2003) note from graphical analysis that concentrations of ammonia-nitrogen appeared to decline in the reservoirs, as well as in the streams draining into the reservoirs from the 1980s through the 1990s. They also noted that in-lake nitrate-nitrogen concentrations appeared to increase in Liberty Reservoir, as well as in its watershed tributaries, but that despite the apparent increase in nitrate-nitrogen concentrations noted in the Loch Raven and Prettyboy Reservoir watershed tributaries, no clearly discernible and consistent in-lake trends in nitrate concentrations were apparent in either reservoir.

Although no statistical analyses were conducted for ammonia- and nitrate-nitrogen concentrations in Liberty Reservoir and its watershed tributaries, trend analyses were conducted for Loch Raven and Prettyboy Reservoirs and their watershed tributaries (Maryland Department of the Environment, 2003d). Ammonia-nitrogen concentrations did significantly decline in dry-weather flows in most Loch Raven and Prettyboy Reservoir watershed tributaries, but significant declines in ammonia-nitrogen concentrations occurred at only two stations (fig. 3B, Station GUN0171 and GUN0156) in the middle-to-lower part of Loch Raven Reservoir (Kendall trend test, with autocorrelation, p less than 0.05, with data from 1981–95). Statistical analyses also revealed positive significant trends in nitrate-nitrogen concentrations in dry-weather flows in the Loch Raven and Prettyboy Reservoir watershed tributaries, but no significant trends in nitrate-nitrogen concentrations were observed in either reservoir.

The lack of a direct correspondence between observed patterns, and in at least some cases, verified trends, in ammonia- or nitrate-nitrogen concentrations in watershed tributaries and reservoirs is not unexpected. First and foremost, the patterns observed in the tributaries only reflect available forms of nitrogen in dry-weather flows. Measurements of ammonia- and nitrate-nitrogen in the reservoirs, however, could reflect not only available nitrogen from dry-weather flows but also stormflows. Although the latter would be expected to have similar or lower concentrations than dry-weather flows, they nevertheless would increase available nitrogen loads to the reservoirs. Reservoir concentrations also could reflect nitrogen made available from organic sources in streamflows (not measured) or within the reservoir through in-lake abiotic and biotic, and (or) hydrodynamic processes—for example, through the release of ammonia-nitrogen from bottom sediments under hypoxic and stratified conditions.

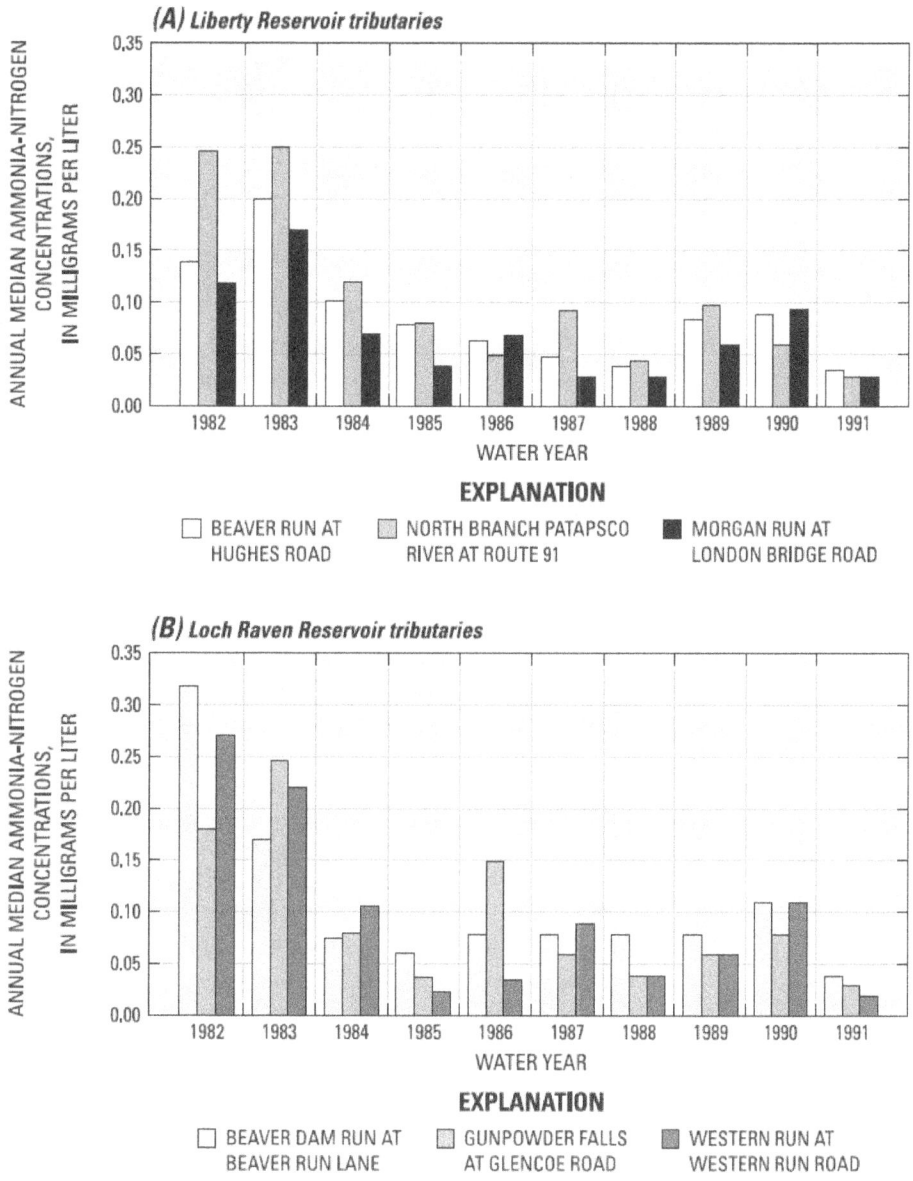

Figure 5. Annual median ammonia-nitrogen concentrations in dry-weather flows at *(A)* Liberty and *(B)* Loch Raven Reservoir tributaries, 1982–91 (modified from Reservoir Watershed Protection Subcommittee, 1992).

Reservoir processes likely do play a role in the determination of in-lake available nitrogen concentrations. For example, monitoring data indicate both ammonia- and nitrate-nitrogen concentrations are considerably lower in the reservoirs than in streams (table 5), an indication of considerable in-lake nitrogen processing and recycling. In addition, concentrations of available forms of inorganic nitrogen appear to be governed in part by reservoir hydrodynamics. During lake stratification, they generally are stratified. Nitrate nitrogen is highest in shallow oxic waters, whereas ammonia nitrogen is highest in deep hypoxic waters (Baltimore City Department of Public Works, 1996). Additional evidence that in-lake processes influence concentrations of available nitrogen is seasonally apparent. As noted earlier, nitrate-nitrogen concentrations in most tributaries tend to peak in the winter months and then gradually decline to their lowest values during the summer months. Concentrations of nitrate nitrogen in the reservoirs, however, tend to peak in the late spring to early summer and then decline (for example, fig. 7), presumably because of midsummer in-lake primary production.

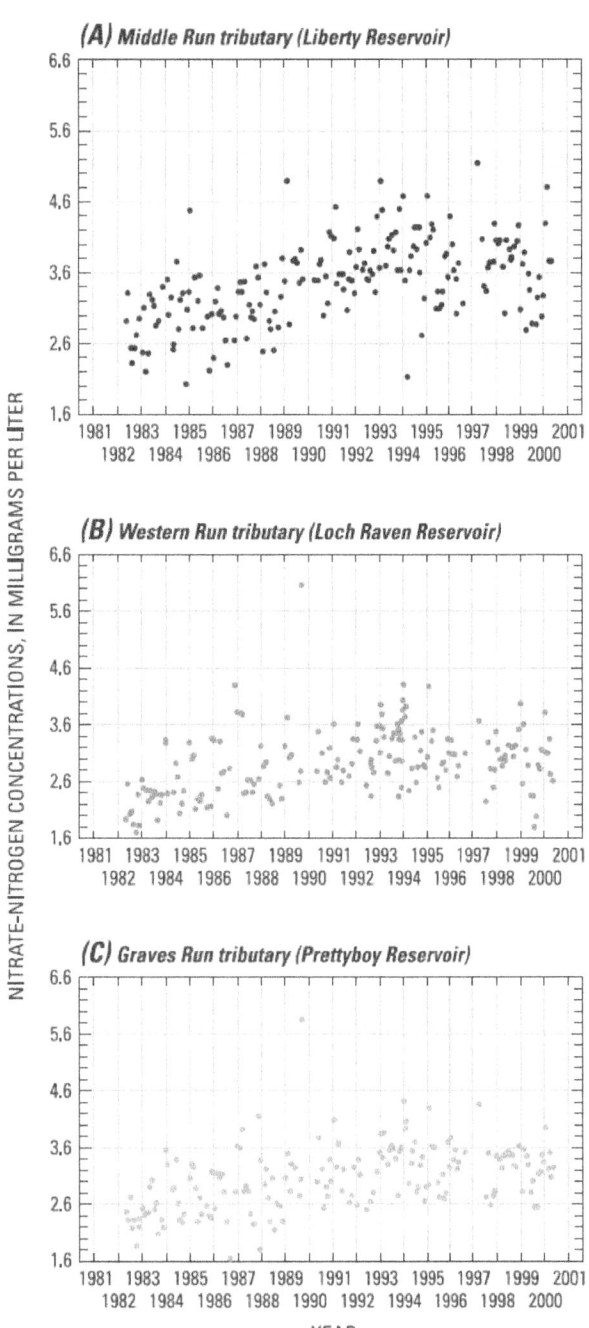

Figure 6. Nitrate-nitrogen concentrations in dry-weather flows at *(A)* Middle Run tributary in the Liberty Reservoir watershed, *(B)* Western Run tributary in the Loch Raven Reservoir watershed, and *(C)* Graves Run tributary in the Prettyboy Reservoir watershed, 1982–2000 (modified from Baltimore City Department of Public Works, 2001).

Phosphorus Transport and Reservoir Cycling

Considerable attention and effort have been paid to the reduction of phosphorus loads to the Baltimore City reservoirs. Amatayakul and others (1978) provided the initial laboratory bioassay studies that led them to conclude that stoichiometric increases in phosphorus rather than nitrogen produced the greatest increase in algal production in euphotic water samples from Loch Raven Reservoir (table 3). Monitoring data also have been used by the RTG and others as evidence from the field that phosphorus possibly is the limiting nutrient (Amatayakul and others, 1978; Interstate Commission on the Potomac River Basin, 2006). Various ratios of nitrogen to phosphorus (N:Ps) have been calculated and reported (10-to-100 or more to one) for shallow surface waters in the Loch Raven and Prettyboy Reservoirs, and have been reported to typically exceed the Redfield (16:1) ratio (Redfield, 1958), the ratio of total nitrogen to total phosphorus found in phytoplankton when neither nutrient is limiting. The reported ratios, however, utilize available nitrogen, or modeled estimates of total nitrogen, rather than actual measurements of total nitrogen. Hence, the actual total N: total P ratios in reservoir surface waters could differ from the reported ratios. In addition, the Redfield ratio was developed from marine not artificial freshwater (reservoir) environments, and the variations in this ratio could be more important to consider than a single ratio value when assessing nutrient limitations and phytoplankton production in fresh waters (Downing and others, 2001; Arrigo, 2005).

Amatayakul, Defries, and others (1978) also provided and compared estimates for TP, with the earliest load estimates for Loch Raven Reservoir (table 4). As is the case with nitrogen loads described earlier (see **Eutrophication—Nutrients: Nitrogen Transport and Reservoir Cycling**, *this report*), they noted a lack of spatial and temporal coverage in the TP data required to estimate loads. Their monitoring recommendations for phosphorus included year-round data collection in the reservoir watersheds for point sources (for example, WWTPs) and in major tributary storm- and base flows to the reservoir.

The long-term reservoir watershed monitoring network created in the early 1980s included monitoring at known major point sources for phosphorus (NPDES WWTPs), in the reservoir watershed tributaries during storm- and base flows, and in the reservoirs. Analyses for phosphorus initially included TP at all monitoring sites, and through the early 1990s, additional analyses for DOP for storm- and selected dry-weather flows at the three gaged tributary stations in each supply reservoir. The measurement of DOP was used to approximate bioavailable phosphorus. No DOP measurements were routinely obtained in the reservoirs, mainly because TP concentrations in the reservoirs often approached or were below the laboratory method reporting levels at the water-treatment facilities (Robert McAuley, Baltimore City Department of Public Works, written commun., 2008). Thus, long-term tracking of phosphorus has mainly relied on TP data alone from dry-weather flows to describe concentrations and (or) compute areal-weighted loads

Table 5. Median ammonia- and nitrate-nitrogen concentrations in the Baltimore reservoirs, 1981–93 (adapted from Baltimore City Department of Public Works, 1996).

[mg/L, milligrams per liter]

Nutrient	Time period	Reservoir location	Reservoir waters[1]	Median concentration (mg/L)		
				Liberty	Loch Raven	Prettyboy
Ammonia-nitrogen	Year-round	Lower	Shallow	0.04	0.05	0.04
	Growing season	Lower	Shallow	0.03	0.03	0.04
	Growing season	Headwaters	Shallow	0.05	0.04	0.03
	Year-round	Lower	Deep	0.06	0.11	0.05
Nitrate-nitrogen	Year-round	Lower	Shallow	1.53	1.43	1.69
	Growing season	Lower	Shallow	1.81	1.50	1.74
	Growing season	Headwaters	Shallow	2.04	1.68	1.78
	Year-round	Lower	Deep	0.14	0.11	0.05

[1] Shallow waters are 30 feet deep or less; deep waters exceed 30 feet in depth.

at all dry-weather-flow tributary stations, or from storm- and dry-weather flows to describe concentrations or total loads, or compute flow-concentration curves, at each of the three gaged tributary stations in the supply-reservoir watersheds.

After WWTPs were identified early on as likely sources of phosphorus that contributed to excessive algal blooms (Valcik, 1975), upgrades to the WWTPs were made during the late 1970s through mid-1980s. Data provided by the long-term monitoring network through the 1980s to mid-1990s were used to document substantial reductions in TP effluent loads following WWTP renovations in the late 1970s through 1980s (Reservoir Watershed Protection Committee, 2000, fig. 12). Summary and other reports (Baltimore City Department of Public Works, 1996; Winfield and Sakai, 2003) indicate that within the Liberty Reservoir watershed, total annual loads of TP in WWTP effluents decreased 95 percent to approximately 28 kg/yr (kilograms per year) for 1989–94, from approximately 512 kg/yr for 1983–86. In the Loch Raven and Prettyboy Reservoir watersheds, the WWTP effluent loads for TP decreased 71 percent during the same period.

Annual TP loads from WWTPs are a small part of the total annual watershed tributary TP loads to the reservoirs. For example, the TP loads from the WWTPs to the streams in Loch Raven Reservoir watershed in what was then considered a wet water year (1984) represented only approximately 3 percent, and in a dry water year (1986), only approximately 9 percent, of the estimated total annual tributary load of TP from the three gaged subbasins to the reservoir in each of these water years (Baltimore City Department of Public Works, 1996, fig. 10.1). Nevertheless, reductions in effluent TP loads were considered important (Baltimore City Department of Public Works, 1996, 2000, 2001). The WWTP effluents contained available forms of nitrogen—ammonia, nitrite, and nitrate—and possibly contained high concentrations of DOP (Baltimore City Department of Public Works, 1996). The

documented reductions of TP in WWTP effluents also demonstrated that the long-term monitoring program effectively can track changes in known point-source TP load through the NPDES program.

Given that effluent loads from suspected major point sources accounted for only a small fraction of the total annual tributary loads for TP, it was suspected by the RTG that the major portion of tributary loads was largely from unregulated nonpoint sources. Thus, the RTG and others periodically have used long-term monitoring data to estimate TP loads to the reservoirs (table 4); the estimated loads are generally greater than the loads initially estimated by Amatayakul, Defries, and others (1978). It should be noted, however, that all TP load estimates, regardless of when they were computed, reflect TP monitoring data obtained before the year 2000, which will be discussed further below.

Monitoring data for TP from the early 1980s to mid-1990s have been used to assess annual variations and differences in TP concentrations and loads among and from the six gaged tributaries and their subbasins (figs. 8A and B; Reservoir Watershed Protection Subcommittee, 1992). Summary reports for this period note that mean total annual TP concentrations and loads varied for each subbasin (Baltimore City Department of Public Works, 1992, 1996; Reservoir Watershed Protection Subcommittee, 1992). The reports also noted that the high variability in TP loads was likely the result of climatic variation—concentrations and loads being notably higher in wet years than in dry years (fig. 9, for example, 1984 and 1986, respectively), and, within a given year, to flow regime. Most of the TP load in the gaged tributary subbasins did indeed appear to be carried by storm-, as opposed to dry-weather, flows, with the latter, depending on the tributary, accounting for no more than 15–30 percent of the total annual TP load (fig. 8B).

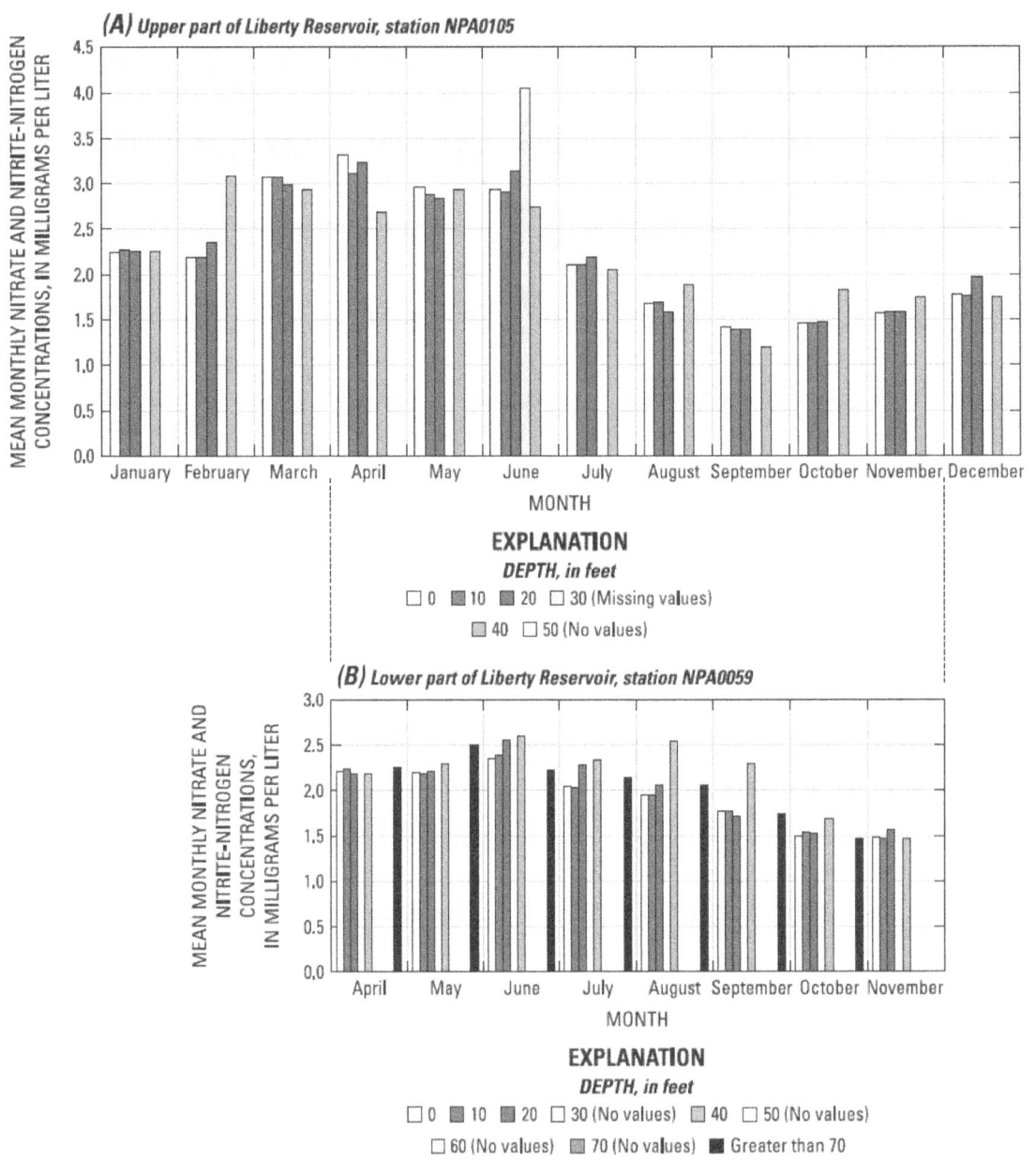

Figure 7.   Mean monthly nitrate- and nitrite-nitrogen concentrations in *(A)* upper (Reservoir station NPA0105), and *(B)* lower (Reservoir station NPA0059) parts of Liberty Reservoir, 1994–2001 (modified from Winfield and Sakai, 2003).

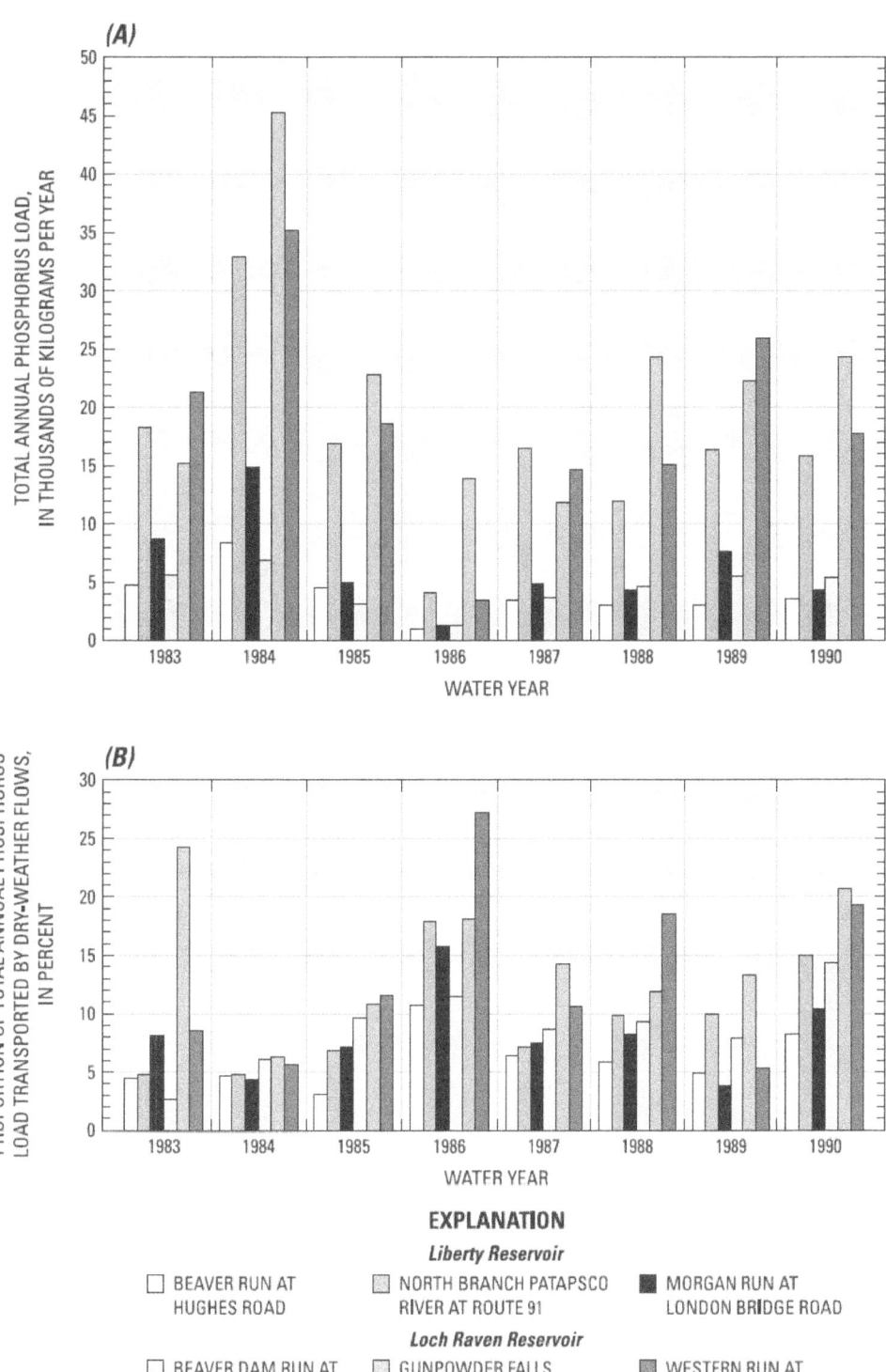

Figure 8.   *(A)* Annual loads of total phosphorus and *(B)* proportion of total annual phosphorus load transported by dry-weather flows for six gaged streams and subbasins in the Loch Raven and Liberty Reservoir watersheds, 1983–90 (modified from Reservoir Watershed Protection Subcommittee, 1992, and Baltimore City Department of Public Works, 1996)

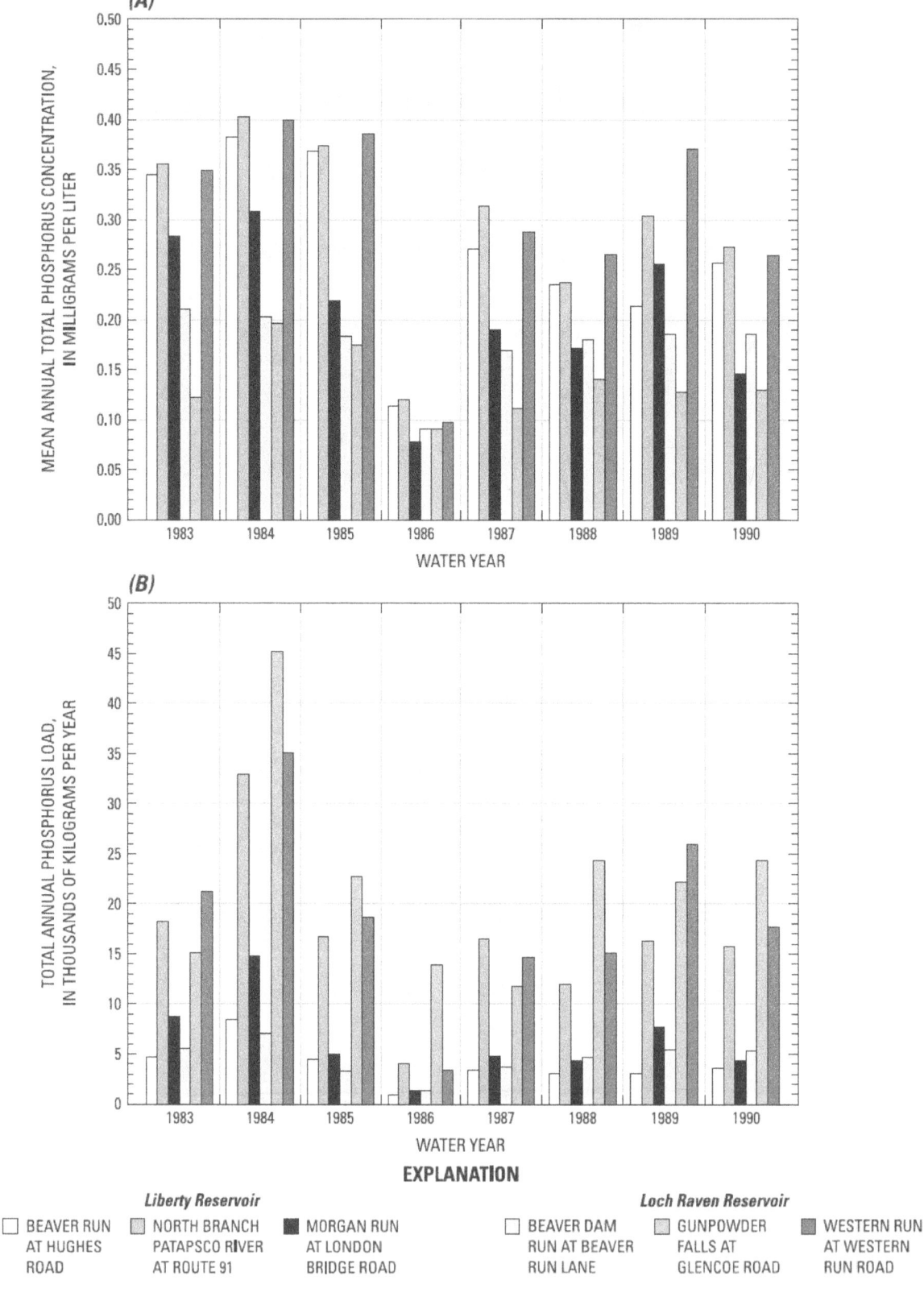

Figure 9. *(A)* Mean annual total phosphorus concentration and *(B)* total annual phosphorus load for gaged streams and subbasins in the Loch Raven and Liberty Reservoir watersheds, 1983–90 (modified from Reservoir Watershed Protection Subcommittee, 1992).

Monitoring data for TP in storm- and dry-weather flows during the 1980s at the six major tributaries in the supply reservoir watersheds were used to provide the first spatial load assessments among major subbasins. On the basis of areal-weighted TP (storm- and dry-weather flow) loads or yields (Baltimore City Department of Public Works, 1996, table 10.1), the three major subbasins within each supply reservoir could be ranked to help identify high TP source areas for management of this nutrient (Baltimore City Department of Public Works, 1996, 2000, 2001). Although not statistically confirmed in these reports, for the Loch Raven Reservoir watershed, Western Run appeared to have the highest yield among the three tributaries, despite the variability yields, which appeared similar among the three tributaries. Apparent differences in yields among the three major Liberty Reservoir watershed subbasins were considered relatively minor and insufficient to warrant ranking given the high variability in yields at all three subbasins.

The long-term monitoring data also have enabled subbasin TP yield comparisons among all reservoir watersheds to be computed, but solely with dry-weather-flow data (Baltimore City Department of Public Works, 1996, fig 9.5). Western Run again appeared to have the highest yields among all monitored subbasins in the Loch Raven Reservoir watershed, as did Georges Run in the Prettyboy Reservoir watershed, and Bonds Run and the North Branch of the Patapsco River at Route 91 in the Liberty Reservoir watershed. Overall, dry-weather-flow TP yields among subbasins, however, also were highly variable, and average yields among subbasins within a reservoir watershed differed by less than 5 to 10 kg/m^2/yr (kilograms per square meter per year) (Baltimore City Department of Public Works, 1996, table 10.1). Whether or not these differences in yields were statistically different among subbasins was not determined.

Dry-weather flow TP data also have been used on a limited basis to assess apparent relations between tributary TP concentrations and subbasin land use and cover characteristics—forested, agricultural, or residential (Baltimore City Department of Public Works, 2001, fig. 27, fig. 30, and fig. 38). Median TP concentrations for dry-weather flows in 1999 for each subbasin were found to be inversely related to the proportion of the subbasin in forest cover. (Presumably this relation holds true for most other years as well.) Median TP concentrations in 1990 and 1999 for each subbasin also appeared to be positively related to the proportion of the subbasin in agricultural land use, and the difference in median TP concentrations from 1990 and 1999 appeared to be positively related to the number of low-density residential housing units without sewer hook-up per square mile of subbasin constructed during the 1990s. The latter relation was taken by the RTG to indicate that construction operations (for example, cut and fill, and septic-system installation) associated with new residential units could result in sediment transport or have some other type of transient effect on TP concentrations in streams. Although the above apparent TP and land-use relations did not consider stormflow, it was inferred they could

be considered in the development of management activities to reduce phosphorus loads in selected individual subbasins (Reservoir Watershed Protection Committee, 2000).

Tributary dry-weather-flow data for TP have been analyzed to assess annual trends and seasonal variations. Summary and other reports (Baltimore City Department of Public Works, 1992, 1996, 2001; Winfield and Sakai, 2003) describe apparent decreases in TP concentrations at most dry-weather tributary stations. The apparent decline in TP in the Liberty Reservoir watershed tributaries began in the 1980s and continued through the 1990s—for example, at Little Morgan and Middle Runs (fig. 10). Similar apparent trends were noted for tributaries in the Prettyboy Reservoir watershed—for example, Gunpowder Falls and Georges Run (fig. 11). Declines in TP concentrations were not apparent for tributaries flowing into Loch Raven Reservoir—for example, at three stations on the Gunpowder Falls (fig. 12).

The City (Baltimore City Department of Public Works, 2001) noted that the apparent declines in TP concentrations for dry-weather flows in Liberty Reservoir and Prettyboy watersheds (fig. 10, fig. 11) appeared to be accompanied by apparent declines observed in algal counts and, to a less apparent degree, concentrations of chl-a, in the reservoirs (fig. 13). In addition, despite no apparent declines in TP concentrations in tributaries in Loch Raven watershed, there were apparent declines in total algal counts and concentrations of chl-a in the reservoir (fig. 13).

To substantiate whether there were significant trends in TP, at least in relation to the Loch Raven Reservoir watershed, the Maryland Department of the Environment (2003d) conducted statistical trend analysis on the tributary dry-weather-flow and reservoir in-lake TP data collected from 1983–92. Their results revealed that among the tributaries, only two stations (Dulaney Valley Branch and Gunpowder Falls at Falls Road just below Prettyboy Reservoir) showed significant negative trends whereas the remaining four tributary stations indicated no significant trend in TP concentrations. In the reservoir, only one in-lake station in the lower part of the reservoir (fig. 3B, Reservoir station GUN0142) showed a significant decline in TP. All other reservoir stations (four) had no significant trends. Overall, the investigation concluded that for the period of record examined, there generally were no significant declines in TP concentrations at most reservoir and tributary stations. Thus, changes in TP concentrations, at least on the basis of dry-weather-flow data, could not account for the apparent long-term declines in reservoir total algal counts or concentrations of chl-a.

Assuming that the apparent declines in algal counts and chl-a concentrations in the reservoirs from 1981–2001 could be statistically verified in all three reservoirs, which is necessary, they could reflect trends in environmental factors other than TP that influence phytoplankton production. For example, if phosphorus is not truly solely limiting phytoplankton production, the declines in ammonia nitrogen in the reservoir tributaries described earlier could have led, or helped lead, to the declines in algal counts and chl-a concentrations

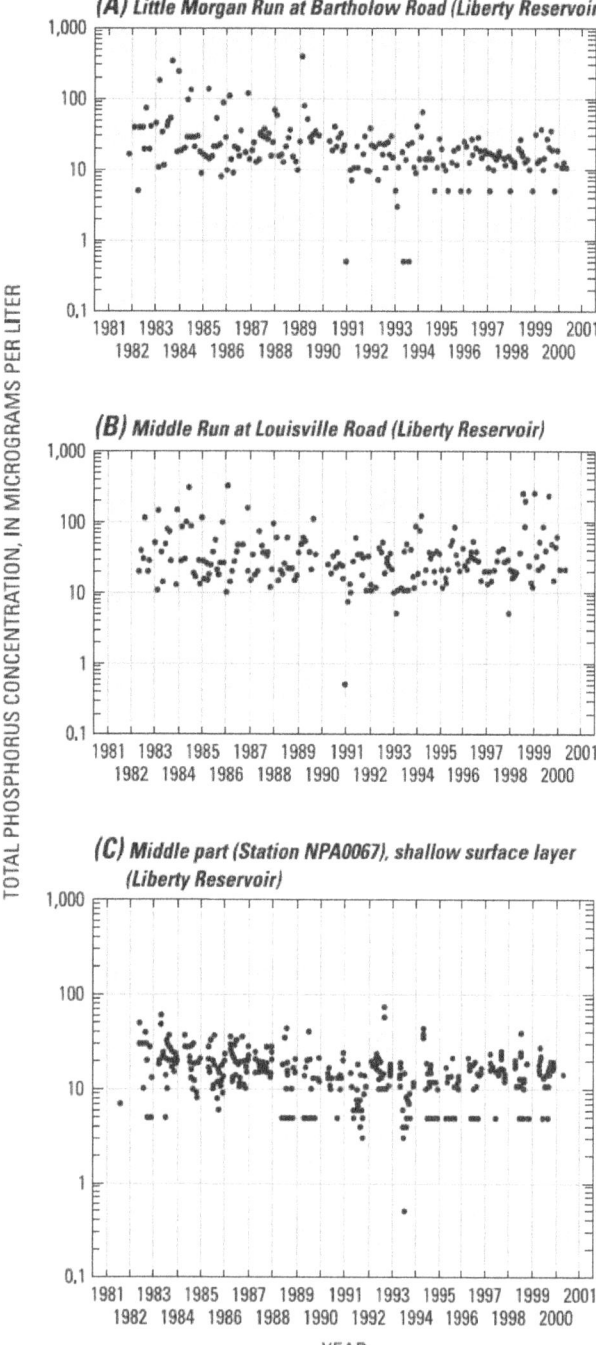

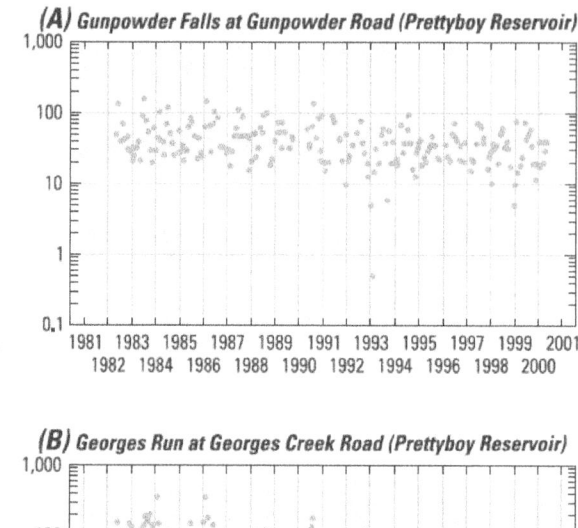

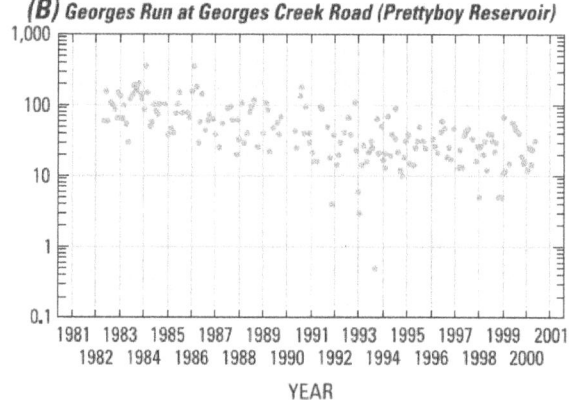

Figure 11. Total phosphorus concentrations for dry-weather flows at selected subbasin tributaries in the Prettyboy Reservoir, 1982–2000 (modified from Baltimore City Department of Public Works, 2001).

Figure 10. Total phosphorus concentrations for dry-weather flows at selected subbasin tributaries in the Liberty Reservoir watershed and in the shallow surface (less than 30-foot depth) layer in the middle part of Liberty Reservoir, 1982–2000 (modified from Baltimore City Department of Public Works, 2001).

in the reservoirs (see **Eutrophication—Nutrients: Nitrogen Transport and Reservoir Cycling**, *this report*). Other environmental factors could influence phytoplankton production, but possibly were not fully investigated. For example, there is evidence of reductions in suspended sediment (a potential source of phosphorus) in tributary flows (see **Sedimentation—Sediment Transport**, *this report*).

Perhaps as important, studies conducted after the investigation by MDE indicate that problems in the monitoring methods for phosphorus could play a role in the inability to explain the differences in apparent trends in TP in the watershed tributaries and for tributaries and in-lake concentrations of TP, chl-*a*, and algal counts. Among the monitoring issues raised are the quality of the TP data, the reliance on measurements of only dry-weather-flow TP concentrations (instead of storm- and base-flow TP concentrations and loads), and the measurement of only TP (rather than some form of phosphorus that could possibly better represent bioavailable phosphorus, for example, DOP, or total dissolved phosphorus, TDP, measured as orthophosphate phosphorus in a filtered and

persulfate-digested sample) (Baltimore City Department of
Public Works, 1992, 1996, 2001; Winfield and Sakai, 2003;
Interstate Commission on the Potomac River Basin, 2006). In
addition, there is a lack of information on phosphorus cycling
in the reservoirs, including in-lake sources of bioavailable
phosphorus possibly released from sediment, and (or) from
early season algal bloom die-offs.

Notable among the concerns raised regarding the limitations in phosphorus monitoring is the quality of the TP data
obtained through the long-term monitoring network. Shortly
after the completion of the Maryland Department of the
Environment (2003d) study, KCI Technologies, Inc. (KCI
Technologies, Inc., 2004) presented its findings on an examination of trends in TP for all three reservoirs. Using monitoring data obtained from 1981 to 2003, they examined the extent
to which non-systematic variations, such as step changes or
outliers, influenced the ability to detect trends in TP concentrations in the reservoirs and their tributaries. They found that all
or almost all stations in each reservoir and reservoir watershed
would exhibit statistically significant declines in TP during all
or most of the period of record, except for two unusual anomalies in the TP data. The first anomaly was an inexplicable step
change (increase) in TP concentrations that occurred in June–
July 1995. The step changes were most pronounced in the
TP data for monitoring stations in Loch Raven and Prettyboy
Reservoirs, and ranged from approximately 40–100 µg/L. The
pronounced step changes were suspected of being the result
of an undocumented change in the analytical procedures at
the Montebello water-treatment laboratory. The second data
anomaly was the occurrence of extreme (generally low) TP
values in the data record. At least some of these low values
were suspected of being TP concentrations that were less than
the laboratory reporting levels (LRLs) for the Montebello and
Ashburton treatment-facility laboratories, which historically
were stored in the monitoring database as real number values
corresponding to one half the LRL with no additional remarks.

Other studies have questioned the quality of the TP
data possibly because of one or both of the above anomalies
(Walker, 1988; Interstate Commission on the Potomac River
Basin, 2006). The lack of sufficient field and laboratory QAC
information and data in relation to laboratory methods used
and possible changes in methods, and the manner in which
concentration data below the LRL historically were stored,
again were cited as shortcomings in the data-collection program. As of 2007, neither of these issues appears to have been
fully and adequately addressed.

From an environmental perspective, limitations in monitoring phosphorus solely as TP can be illustrated by results
obtained by the RTG to characterize seasonal variations in TP
in the tributaries and reservoirs. Concentrations of TP in the
reservoir watershed tributaries in dry-weather flows have been
shown to vary seasonally. Peak intra-annual (monthly) median
concentrations of TP appear to occur in the summer months
and are at least 2 to 5 times greater than median monthly TP
concentrations during most of the rest of the year (for example, Loch Raven watershed, fig. 14).

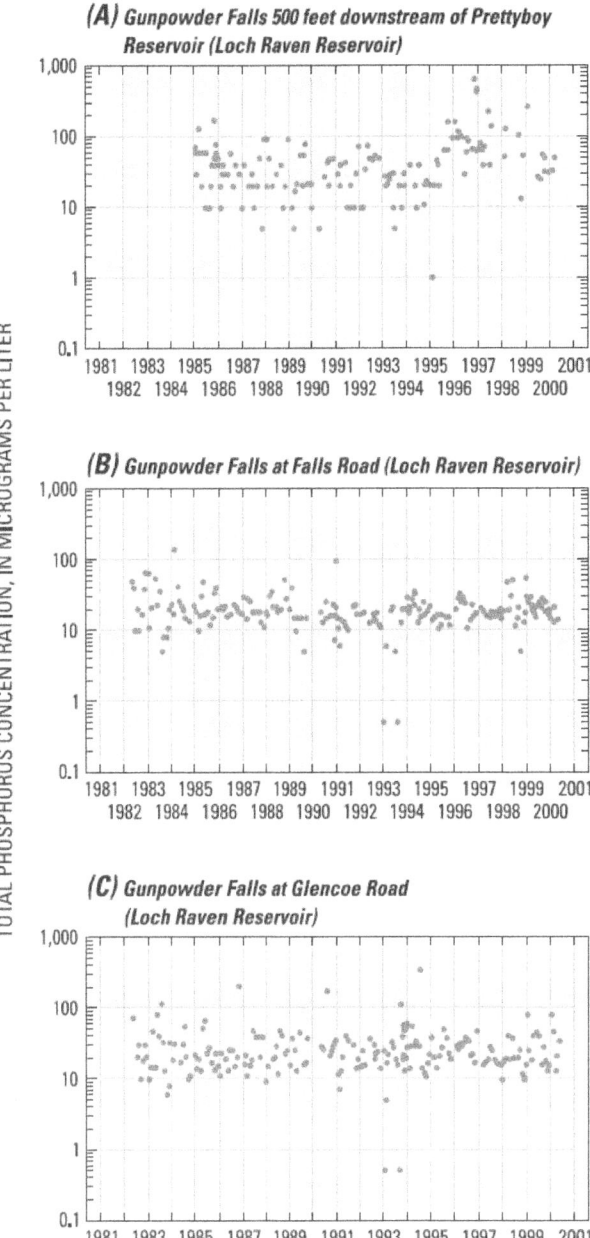

Figure 12. Total phosphorus concentrations for dry-weather
flows in the Loch Raven Reservoir watershed along Gunpowder
Falls, from just below the Prettyboy Reservoir outlet to Loch
Raven Reservoir, 1982–2000 (modified from Baltimore City
Department of Public Works, 2001).

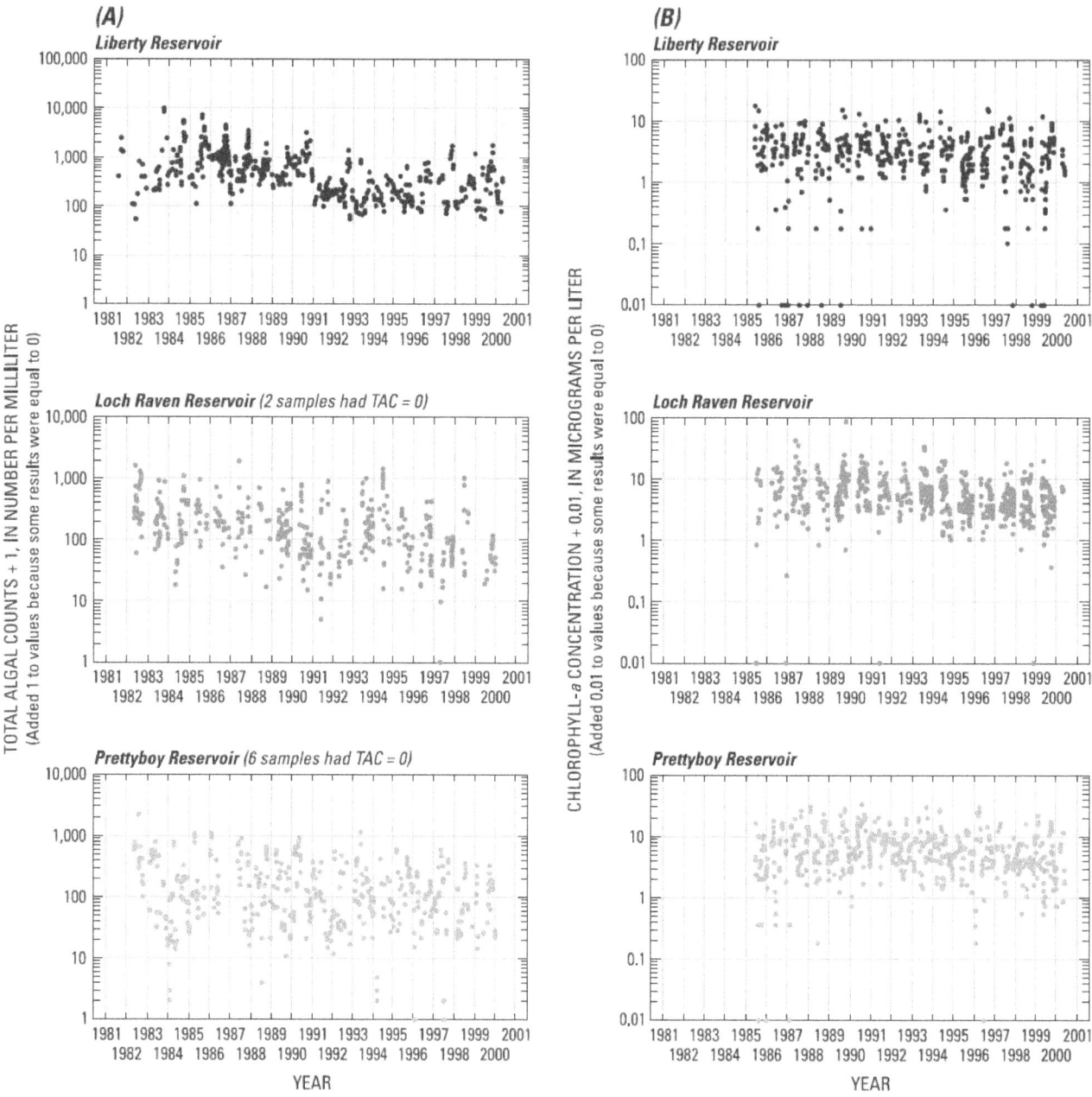

Figure 13. *(A)* Total algal counts (TAC) and *(B)* chlorophyll-*a* concentrations for shallow surface waters (less than 30 feet deep) at selected stations in the Liberty, Loch Raven, and Prettyboy Reservoirs, 1981–2001 (modified from Baltimore City Department of Public Works, 2001).

The elevated concentrations of TP in summer flows have been cited as a concern possibly warranting further investigation (Baltimore City Department of Public Works, 1996, 2001). Their arrival does occur just before most major mid-to-late summer algal blooms begin in the supply reservoirs, and when epilimnetic conditions such as light intensity and penetration and water temperature are conducive to algal blooms. Because of these conditions, it was noted in the above reports that identification of subbasins with the highest peak summer TP concentrations possibly could be used to help focus management activities in these subbasins to reduce summer TP inflows, and thus possibly reduce mid-to-late-summer phytoplankton production.

Whether or not the summer peaks in dry-weather-flow TP concentrations do contribute to mid-to-late-summer algal blooms is uncertain. They could possibly act as a tipping point, or add to the available phosphorus pool and possibly exacerbate algal blooms, since their arrival does coincide with a period when epilimnetic water temperatures and microbial activity are elevated. But peak summer TP dry-weather-flow concentrations correspond to low flows, and thus loads. For example, areal-weighted median or average summer dry-weather flows (at least based on measured values at the time of dry-weather sampling) only are on the order of a cubic foot per second per square mile (fig. 14), or, in terms of stream-flows, in flows of a few tens of cubic feet per second or less.

Adding to the uncertainty of the effect of peak TP dry-weather flows on summer algal blooms is that there is no corresponding TP summer peak observed at most in-lake stations. Reservoir TP concentrations generally are similar in magnitude to dry-weather TP concentrations. If tributary inflows exerted that much influence, some elevation in TP concentrations in reservoir stations closest to tributary inflows would be anticipated. Elevated reservoir TP concentrations in Loch Raven Reservoir, however, only appear to occur in the shallow surface water of one reservoir station (Baltimore City Department of Public Works, 1996, fig. 9.10) at the expected May-June TP peak for tributaries to this reservoir (fig. 14). Elevated TP concentrations do not appear in upper Liberty Reservoir shallow headwaters in August (Winfield and Sakai, 2003, fig. 7-4d), where tributary-inflow effects are most pronounced, and when seasonal summer peaks are observed at the dry-weather-flow tributary in the watershed for this reservoir. Also, in neither case do the tributary dry-weather-flow seasonal patterns in TP to either water-supply reservoir explain the seasonal variations, or annual TP peak concentrations, that occur generally in the fall, in both the headwaters and elsewhere, at most stations in these reservoirs (Baltimore City Department of Public Works, 1996, fig. 9.10; Winfield and Sakai, 2003, fig. 7-14d). These seasonal peaks in the reservoirs could reflect phosphorus release from algal bloom die-offs or sediment, and the beginning of lake turnover. Thus, reservoir abiotic and biotic, and hydrodynamic, processes likely determine, in whole or at least in part, the availability of phosphorus, and concentration of TP, in the reservoirs.

Without measurements of bioavailable phosphorus (for example, DOP or TDP) in the tributaries and reservoirs, however, one cannot determine if the seasonal variations in TP in both the tributaries and reservoirs truly reflect variations in bioavailable phosphorus. Nor can one determine the relative contributions of bioavailable phosphorus in the reservoirs from tributaries and in-lake sources.

The RWMA partners initially did attempt to quantify bioavailable phosphorus by monitoring DOP during the 1980s through the early 1990s. From 1981–93, concentrations of DOP were more or less routinely determined as part of storm sampling, less frequently determined for dry-weather-flow sampling, and as noted earlier, seldom if ever determined in the reservoirs. Monitoring data for TP and DOP at the three gaged watershed tributaries in Liberty Reservoir were summarized by Winfield and Sakai (2003). Long-term average TP and DOP concentrations in tributary stormflows were 0.38–0.42 mg/L and 0.08–0.15 mg/L, respectively, and for dry-weather flows were 0.02–0.05 mg/L and 0.02–0.03 mg/L, respectively (table 6). On average, for the period from 1981–93, and depending on the tributary, only a small fraction of tributary stormflow TP was DOP—approximately 4 to 21 percent; however, approximately 75 to 81 percent of dry-weather-flow TP was DOP.

Relative to the period during which DOP data were collected, there also was a notable change in the ratio of DOP to TP in tributary flows. About mid-way through the 1981—93 monitoring period, there was an abrupt decline in the ratio of annual median DOP concentrations to annual median TP concentrations in gaged tributary flows (fig. 15). What caused the decline in this ratio is not known. The City (Baltimore City Department of Public Works, 1992) thought that the decline in the ratio could reflect a combination of watershed best management practices, including changes in agricultural methods or practices, modifications to WWTPs, and a statewide ban on phosphates in detergents. They also noted, however, that the decline was being investigated to determine if it was due to a bias in the sampling method used during stormflows up through 1986. No further results from this investigation were encountered in any subsequent reports obtained for this review. The decline in the DOP:TP ratio after 1986 implied that the median annual DOP concentration accounted for no more than approximately 40 percent of the median annual TP from gaged tributary storm- and dry-weather flows; before 1987, the median annual DOP concentration had accounted for at least 40 percent or more of the median annual TP concentration at most of the gaged tributary sites.

The combination of finding a relatively high ratio of DOP to TP (DOP:TP) in dry-weather flows from 1980–93, combined with the low ratio of DOP:TP in storms after 1986, possibly contributed to the shift in the focus of the monitoring program in the early 1990s—from monitoring of DOP and TP in storm- and dry-weather flows to monitoring just TP in dry-weather flows as a means of tracking changes in phosphorus in the reservoir watershed tributaries. It also was apparent that dry-weather-flow monitoring of TP, which contained

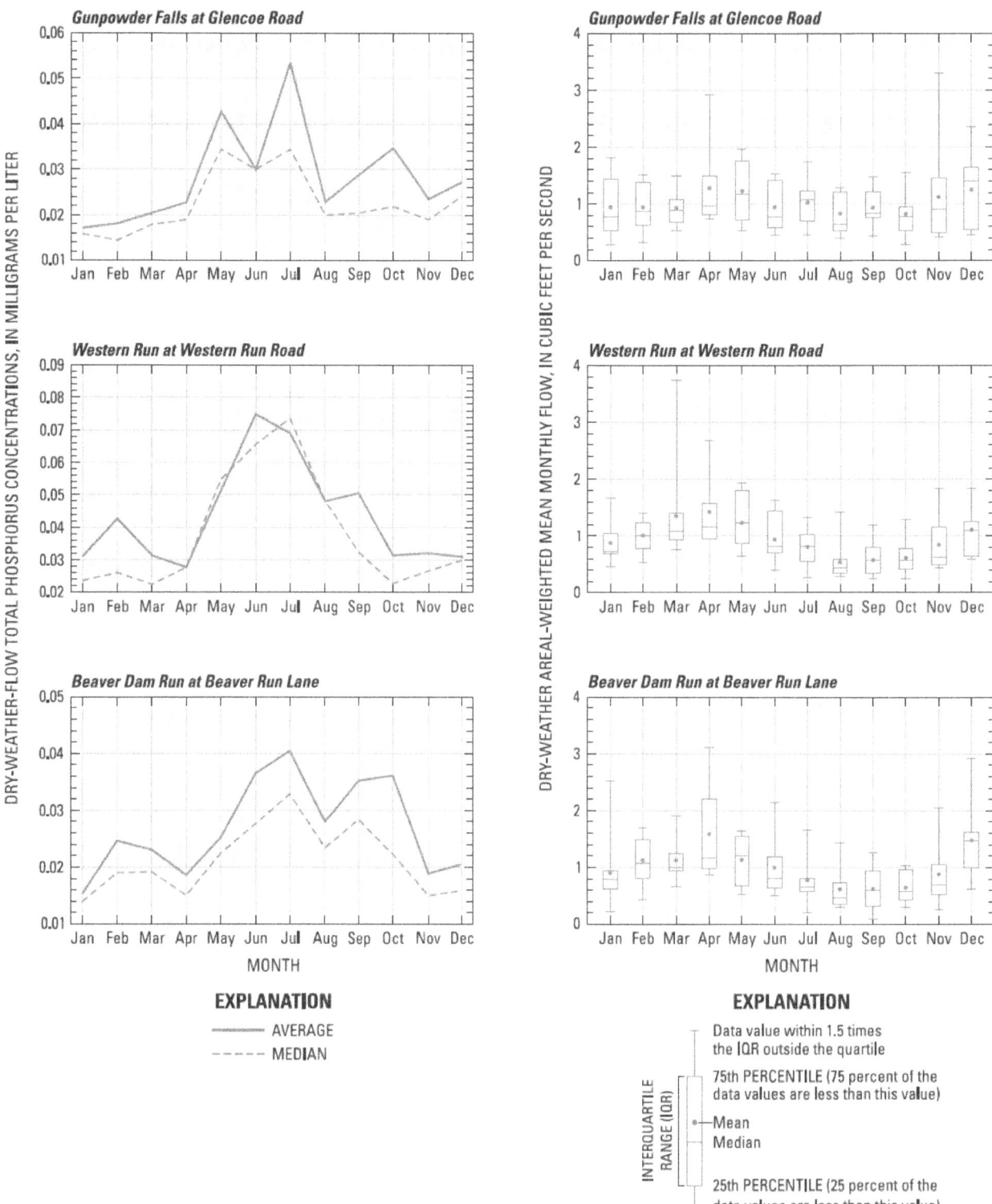

Figure 14.   Seasonal patterns in monthly total phosphorus concentrations and rates of dry-weather flow for tributaries in the Loch Raven Reservoir watershed, 1981–93 (modified from Baltimore City Department of Public Works, 1996).

Table 6. Mean total phosphorus and dissolved (orthophosphate) phosphorus concentrations and yields for Liberty Reservoir watershed tributaries, 1981–93 (modified from Winfield and Sakai, 2003).

[ft³/s, cubic feet per second; mg/L, milligrams per liter; kg/km², kilogram per square kilometer; ---, unavailable]

Station	Number of samples		Tributary discharge		Concentration		Yield	
	Total phosphorus	Dissolved (orthophosphate) phosphorus	Mean (ft³/s)	Maximum (ft³/s)	Total (mg/L)	Dissolved (orthophos-phate) (mg/L)	Total (kg/km²)	Dissolved (orthophos-phate) (kg/km²)
Dry-weather flows								
Beaver Run at Hughes Road	113	27	12	42	0.025	0.020	7.2	6.5
Morgan Run at London Bridge Road	113	30	24	115	0.021	0.017	6.1	4.8
North Branch Patapsco River at Route 91	110	27	41	145	0.045	0.034	15.3	12.0
Stormflows								
Beaver Run at Hughes Road	173	78	91	1, 917	0.382	0.080	---	---
Morgan Run at London Bridge Road	179	80	201	2, 985	2.64	0.097	---	---
North Branch Patapsco River at Route 91	167	79	401	6, 173	3.22	0.152	---	---

considerable DOP, also would provide data for comparisons among more subbasins than TP monitoring for storm- and dry-weather flows (Reservoir Watershed Protection Subcommittee, 1992).

Despite the possible advantages of just dry-weather-flow TP monitoring, the reliance on this type of monitoring, and the reduction in stormflow TP monitoring, possibly have limited the ability of the RTG to address RWMA concerns. Dry-weather-flow monitoring presents a challenge in formally defining what constitutes a dry-weather flow—to date, these flows have not been well-defined and described in any RWMA partner reports beyond 1993 (fig. 14). More importantly, monitoring for only TP in dry-weather flows could lead to erroneous conclusions regarding changes or comparisons in TP and thus presumably DOP among subbasins if the DOP:TP ratio changes markedly for one or more subbasins through time. Evidence presented above indicates this ratio changed considerably in the mid-1980s, and for reasons that have never been fully understood. Given that no DOP data have been routinely collected for the watershed tributaries since the early 1990s, the DOP:TP ratios are unknown for the watershed tributaries after 1986, and thus after nearly two additional decades

(1990–2007) of efforts to reduce phosphorus loads from the watersheds.

Ultimately, if the main concern with phosphorus in the tributaries is DOP rather than TP, the mean annual concentration of DOP in stormflows is at least five times greater than the mean annual concentration of DOP in dry weather flows (table 6). Thus, the major portion of the DOP load from the tributaries to the water-supply reservoirs in all but the driest years likely occurs from storm- and not dry-weather flows, and the frequency and timing of those storms could have considerable influence on reservoir phytoplankton production.

There have been other major limitations introduced by the reduction in monitoring storms in the mid-1990s. The decline in frequency of storm monitoring after about 1995 (table 7) means tributary TP loads, as well as any other type of load estimates, to the reservoirs in relation to storm- and dry-weather flows will be a challenge to estimate for any reservoir tributary. In turn, models designed to estimate phosphorus loads to either supply reservoir could have too few storm events since 1995 for calibration and validation purposes—for example, see Interstate Commission on the Potomac River Basin (2006).

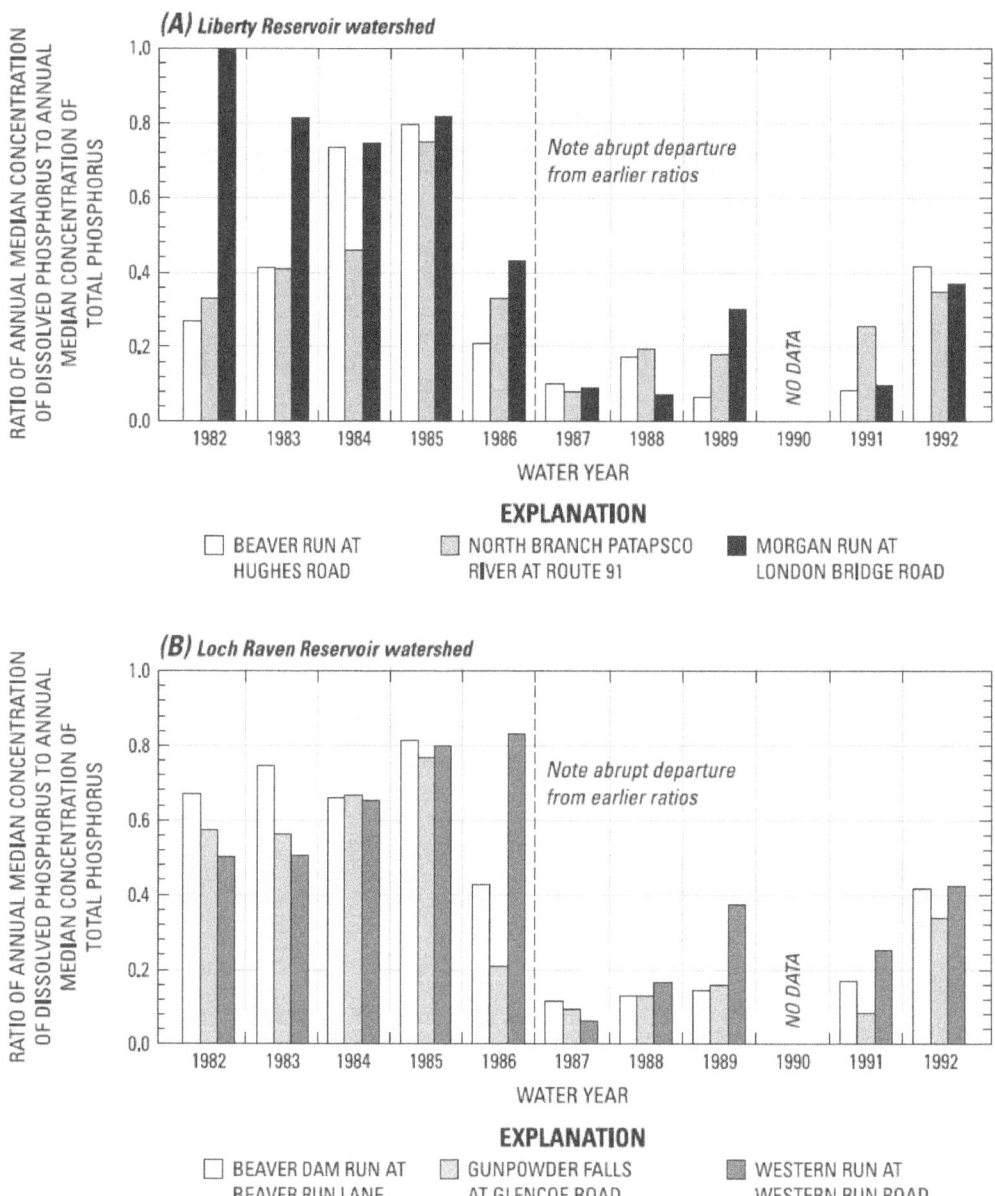

Figure 15.   Ratio of annual median dissolved phosphorus concentration to annual median total phosphorus concentration in gaged tributaries of the *(A)* Liberty and *(B)* Loch Raven Reservoir watersheds, 1982–92 (modified from Baltimore City Department of Public Works, 1992).

Table 7. Number of storm events with water-quality sampling, Loch Raven and Liberty Reservoir watersheds, 1994–2008 (data courtesy of Baltimore City Department of Public Works, Baltimore, Maryland).

Reservoir/ watershed	Water year							
	1994	1995	1996	1997	1998	1999	2000	
Loch Raven	5	0	0	0	4	4	0	
Liberty	5	0	0	0	0	0	3	

Reservoir/ watershed	Water year							
	2001	2002	2003	2004	2005	2006	2007	2008
Loch Raven	3	1	3	3	3	4	3	2
Liberty	2	0	0	1	4	1	4	3

To illustrate the potential importance of modeling to the ultimate goals of the Baltimore Reservoir Program, and the need for adequate data in order to estimate loads, one can examine the results from several chronological studies that at least in part aimed to estimate phosphorus loads for the Baltimore reservoirs (table 4). The most recent annual load estimates (Maryland Department of the Environment, 2003d; Interstate Commission on the Potomac River Basin, 2006) are 4 to 5 times greater than the earliest estimated loads. The most recent TP load estimates relied on more advanced models, but calibration of those models has been based on storm- and dry-weather flow data mostly collected before the latter part of the 1990s, as storm monitoring was discontinued in the mid-1990s, and has only partially been restored since the year 2000 (table 7). Furthermore, in the case of Prettyboy Reservoir watershed, all load estimates are truly estimates as no stormflow data have ever been obtained to at least partially validate estimated loads for the three major tributaries that supply this reservoir. Despite these limitations, accurate load estimates minimally are necessary to help establish suspended-sediment and phosphorus TMDLs for Liberty Reservoir (see **Regulatory Concerns**, *this report*).

In addition to the reduction in monitoring TP in storm-flow, the cessation of monitoring DOP in any tributary flows has further limited the ability of the RTG to address RWMA concerns. Despite the long-term decision to rely primarily on TP as the measurement for phosphorus, the importance of bioavailable phosphorus, estimated by either DOP or TDP, to algal production in the reservoirs has been raised by the City DPW and others. Summary and other reports (Baltimore City Department of Public Works, 1992, 1996, 2001; Winfield and Sakai, 2003; Interstate Commission on the Potomac River Basin, 2006) have outlined concerns regarding the role of and possible need to resume measurements of bioavailable phosphorus (for example as DOP or TDP). Use of TP data alone could be misleading in relation to:

a) Identifying high-phosphorus-source subbasins within the watersheds, and within those subbasins, the landscape conditions and human activities that contribute the most bioavailable phosphorus (not just TP) to watershed tributaries, to better target management activities to reduce phosphorus;

b) Describing the states and trends in watershed conditions in relation to bioavailable phosphorus (not just TP) in the watershed tributaries, in order to better relate watershed tributary contributions of bioavailable phosphorus to in-lake bioavailable phosphorus concentrations and phytoplankton production (algal counts and chl-*a* concentrations), and

c) Determining the levels and relative source contributions of not just tributary, but in-lake biotic (for example, algal decomposition) or abiotic (for example, bottom sediment release), contributions to bioavailable phosphorus (not just TP) in the reservoirs to better identify the in-lake phosphorus sources that possibly contribute to the occurrence of major algal blooms.

In addition, the above reports reveal a less than complete and possibly necessary understanding of nutrient cycling—for nitrogen, as well as phosphorus—in the reservoirs, and of the role that nutrient cycling processes in the reservoirs, as well as tributary loads, play in the timing, occurrence, intensity, duration, and type of algal blooms associated with eutrophic compared to mesotrophic conditions.

Trophic State of Reservoirs

The Carlson Trophic State Index, or TSI (Carlson and Simpson, 1996), is a multimetric index that has been used by the RTG to assess eutrophication in the reservoirs. The TSI value of most concern for the RTG is that which marks the change during mid-to-late spring through early fall (approximately April through September) from moderately impaired (mesotrophic) to severely impaired (eutrophic) conditions (Baltimore Reservoir Technical Group, 2004). Threshold values for the Baltimore reservoirs that correspond to this transition have been determined by the RTG for shallow surface-water concentrations of chl-*a*, DO, and TP, and for water transparency using Secchi depth (Winfield and Sakai, 2003). Using these TSI indicators, the Baltimore Reservoir Technical Group (2004) determined that each reservoir exhibits some degree of eutrophication during the mid-spring through early fall. The degree of impairment among the reservoirs differs, however, depending upon the TSI index used (table 8). These differences are examined in relation to the manner of data collection, aggregation, and analysis for each of the TSI indices.

Phytoplankton and Chlorophyll-a

The concentration of chl-*a* is an indicator of the abundance of phytoplankton algae, and appears to be a sensitive indicator of algal productivity. The TSI index threshold concentration used to distinguish eutrophic from mesotrophic conditions in the shallow surface layer for chl-*a* is 10 µg/L. On the basis of this threshold concentration, the Baltimore Reservoir Technical Group (2004) considered Prettyboy Reservoir to have the highest frequency of eutrophic conditions (be the most impaired), followed by Loch

Raven Reservoir, and then Liberty Reservoir (table 8). In addition, and subsequent to the publication of these data, the Baltimore Reservoir Technical Group (2004) also noted that mean monthly chl-*a* concentrations (and total algal counts) appeared to decline at most monitoring stations in all three reservoirs during the 1980s–90s (fig. 13), but increased during 1999–2003, possibly due to drought and recovery conditions. Winfield and Sakai (2003) and the Interstate Commission on the Potomac River Basin (2006) also indicated that differences in chl-*a* (or algal counts) occur not only among the reservoirs, but within a reservoir, over time. They indicated that these differences reflect variations in the individual hydrodynamics, the magnitude, timing, and duration of peak algal growth periods, and the diversity of taxa among the reservoirs.

There are limitations in the manner the RTG has used chl-*a* as a TSI indicator of mesotrophic and eutrophic conditions. These limitations relate to differences in the frequency and location of sampling of chl-*a* among the reservoirs, and the manner in which the resultant data have been aggregated. Sampling for chl-*a* was only conducted monthly, and only in the lower and middle parts of Prettyboy Reservoir, but bimonthly and throughout each supply reservoir, during periods of stratification. It is not known whether these differences in sampling frequency and the extent of sampling for chl-*a* among the reservoirs affected the frequency of occurrence in eutrophic conditions among the reservoirs on the basis of the results for the TSI chl-*a* index (table 8).

In relation to the depths used to define shallow- and deep-water conditions to aggregate chl-*a* data (table 8), the Interstate Commission on the Potomac River Basin (2006) noted that the actual epilimnion for Loch Raven and Prettyboy Reservoirs derived on the basis of reservoir temperature

Table 8. Reservoir trophic conditions during mid to late spring through early fall, 1981–2000 (from Baltimore City Department of Public Works, 2001).

[µg/L, micrograms per liter; mg/L, milligrams per liter; ≥, greater than or equal to; ≤, less than or equal to <, less than; >, higher frequency of impairment than]

Water-quality index, mesotrophic-eutrophic, or other quality threshold	Frequency (in percent) of all samples that indicate eutrophic conditions			Relative order of reservoir impairment	Period of data[2]
	Liberty	Loch Raven	Prettyboy		
Shallow-water[1] chlorophyll-*a*, ≥ 10 µg/L	9.9	19.8	26.2	Prettyboy > Loch Raven > Liberty	1985–2000
Shallow-water total phosphorus, ≥ 26 µg/L	15.7	57.2	55.5	Loch Raven > Prettyboy > Liberty	1982–99
Secchi depth, ≤ 6.10 feet	5.8	7.9	7.3	Loch Raven > Prettyboy > Liberty	1983/84–2000
Shallow-water dissolved oxygen, < 5 mg/L	10.9	15.7	8.2	Loch Raven > Liberty > Prettyboy	1981/82–2000
Deep-water[1] dissolved oxygen, < 1 mg/L	12.0	48.1	24.6	Loch Raven > Prettyboy > Liberty	1982–2000

[1] Shallow water is up to 30 feet deep; deep water (dissolved oxygen) exceeds 30-foot depth.

[2] Secchi-depth and dissolved-oxygen data collection for Prettyboy Reservoir began 1 year later than data collection at other reservoirs.

profile data and simulations of mixing in the reservoirs is approximately 15–20 ft for a number of years of record, not the 30 ft or less used by the City to define the shallow surficial layer of the reservoirs (for example, Baltimore City Department of Public Works, 2001; Winfield and Sakai, 2003). Thus, only historical data from the near surface, 10-ft depths, and on occasion, the 20-ft depths, likely reflect the true epilimnetic layer in each reservoir for a number of years of record. The inclusion of the 20-ft and 30-ft measurements in the shallow-water layer likely biases the chl-*a* trophic summary data in relation to the true epilimnetic and hypolimnetic layers (table 8). The shallow-water layer chl-*a* data systematically include measurements made at 20-ft and 30-ft depths, which exceed the depth of the modeled epilimnion.

Aggregation of TSI chl-*a* data over each entire reservoir and from all years of data collection could mask noteworthy differences in the degree of eutrophic impairment in the actual epilimnion within, as well as among, the reservoirs. For example, the concentrations of chl-*a* (and algal counts) have been shown to be greatest and extend over greatest depths in the upper part of Liberty Reservoir, where tributaries enter the reservoir (Winfield and Sakai, 2003). Thus, the upper part of this reservoir could experience eutrophic conditions much more frequently, and the lower part of this reservoir much less frequently, than the aggregated TSI chl-*a* data for this reservoir imply (table 8).

In relation to analyses conducted, no analyses of the TSI chl-*a* index data were encountered in RTG, City, or other reports on spatial differences in the frequency of occurrence of mesotrophic or eutrophic conditions within a reservoir. Nor were any analyses encountered on annual trends or seasonal variations in the frequency of occurrence of mesotrophic or eutrophic conditions, within and among the reservoirs. Thus, it is not known whether there has been a decline in the frequency of eutrophic conditions throughout or in parts of any reservoir through the 1980s–90s despite an apparent decline in chl-*a* and algal counts during this period (fig.13).

It should be noted that the aforementioned limitations described for the manner in which the TSI chl-*a* index data were collected, aggregated, and (or) analyzed also apply to other reservoir TSI indices. The parameters used in other TSI indices (see **Eutrophication—Trophic State of Reservoirs: Phosphorus, Dissolved Oxygen,** and **Water Transparency**, *this report*) also were only collected on a monthly basis, and only in the lower and middle parts, of Prettyboy Reservoir. The use of the fixed 30-ft or less depth layer to distinguish between shallow and deep reservoir waters for the TSI chl-*a* index also was used to partition monitoring data composed of incremental depth measurements for the TSI-TP and TSI-DO indices. In addition, aggregation of data for each of the three other TSI indices (using TP, DO, and Water Transparency data) was done over an entire reservoir (table 8), which could mask important differences in the frequency of eutrophic conditions in the epilimnion or hypolimnion within a reservoir on the basis of each of these other indices.

Phosphorus

In theory, phosphorus could serve as a sensitive indicator of eutrophic conditions. Reports by the RWMA partners and others have emphasized that it could be the limiting nutrient to phytoplankton productivity (see **Eutrophication—Nutrients: Phosphorus Transport and Reservoir Recycling,** *this report*).

In relation to the Carlson TSI, a TP concentration of 26 µg/L marks the threshold between eutrophic and mesotrophic conditions for the Baltimore reservoirs (table 8). Relative to this threshold, the Baltimore Reservoir Technical Group (2004) found that Loch Raven Reservoir, followed closely by Prettyboy Reservoir, both appear markedly more impaired than Liberty Reservoir. Relative to this finding and assuming the trend analyses for TP corrected for possible laboratory and data management errors is valid, concentrations of total TP have been shown to decline at most tributary and reservoir stations during the 1980s–90s, but most notably in Liberty Reservoir and watershed (see **Eutrophication—Nutrients: Phosphorus Transport and Reservoir Recycling,** *this report* and KCI Technologies, Inc., 2004).

It also was noted by the Baltimore Reservoir Technical Group (2004), and later substantiated by the Interstate Commission on the Potomac River Basin (2006), that estimated tributary TP yields are greater per acre to Liberty Reservoir than to Loch Raven Reservoir. To explain how greater loads could result in less impairment, the Baltimore Reservoir Technical Group (2004) indicated Liberty Reservoir has about twice the volume of the other reservoirs, but only about half the watershed area (see table 1). This implies that Liberty Reservoir provides a greater potential for dilution of tributary inputs, and because of its longitudinal shape, for the settling of any solids (for example, sediment or algal residues associated with TP) than either of the other two reservoirs.

Despite the implications of the TP trend and spatial analyses described above, no analyses of the TSI-TP index data were encountered in RTG, City, or other reports on annual trends or seasonal variations in the frequency of eutrophic or mesotrophic conditions within a reservoir. Nor were any analysis of the TSI-TP data encountered on whether spatial differences occur in the frequency of eutrophic or mesotrophic conditions within a reservoir.

Dissolved Oxygen

Data for DO concentrations, routinely collected as part of the monitoring program, are used in conjunction with the following adopted standards for DO to develop a TSI-DO index to assess reservoir biotic health (Baltimore Reservoir Technical Group, 2004):

a) Shallow surface-layer water maintains a DO concentration of at least 5.0 mg/L, a State water-quality standard considered adequate to support fish life; and

b) Deep-water layer maintains a DO concentration of at least 1.0 mg/L, a RTG water-quality standard, considered adequate and reflective of mesotrophic (rather than eutrophic) lake conditions, although this level is generally not supportive of normal aquatic life.

On the basis of the TDI-DO index, shallow surface waters for Loch Raven Reservoir appear to be the most impaired, shallow surface waters in Liberty Reservoir are at an intermediate level of impairment, and for Prettyboy Reservoir, they are the least impaired (table 8). Deep waters, however, indicate a high degree of impairment compared to shallow waters for Loch Raven and Liberty Reservoirs, which alters the relative order of impairment—Loch Raven Reservoir being the most impaired, Prettyboy intermediate, and Liberty Reservoir the least impaired.

Whereas the use of DO concentrations and TSI-DO index threshold in the manner described above allows for simple comparisons among reservoirs or between shallow- and deep-water layers within a reservoir, the results (table 8) do not fully illustrate differences in DO concentrations and impairments within a reservoir. First and foremost, all DO measurements in the reservoirs are exclusively made during daylight hours. Measured shallow-water DO concentrations likely reflect phytoplankton production, and could be positively biased because of photosynthetic activity. Hence, the frequency of eutrophic conditions derived from the TSI-DO index could be underestimated in the shallow-water layer of all three reservoirs.

From a broad perspective, the analyses of temporal variations (mean monthly) and spatial differences (by location and within location by depth) of actual DO concentrations within each reservoir indicate that the degree of impairment due to low-DO (eutrophic) conditions is considerably more spatially complex than the generalized analysis of the TSI-DO index data indicates (table 8). All three reservoirs appear to exhibit DO impairments on a seasonal basis for most years of record. For example, Winfield and Sakai (2003) describe a temporal pattern in DO concentrations for Liberty Reservoir, which is similar to the patterns observed in Loch Raven and Liberty Reservoirs (Interstate Commission on the Potomac River Basin, 2006, Appendix C). From January through April, Liberty Reservoir waters have fairly homogeneous and elevated DO concentrations (fig. 16). As thermal stratification begins in the spring (roughly May), the DO concentrations, particularly in the deep-water layer, begin to decline. They continue to decline throughout the summer, presumably because of an increase in biological oxygen demand as a result of phytoplankton residue decomposition, and possibly an increase in water temperatures at depth. The decline in DO is most pronounced in the deepest waters, but occasionally extends into shallow surface waters, which Winfield and Sakai (2003) considered included DO measurements at the 20–30-ft depths. Minimum DO concentrations frequently occurred in the late summer to mid fall (generally August to November). Also during this period, episodic cooling of reservoir waters resulting from variably autumnal weather can introduce episodic thermal mixing, which eventually evolves to persistent cooler temperatures and turnover to re-establish elevated and relatively uniform DO concentrations throughout the reservoir. (Although evidence of this episodic cooling before turnover commences has not been fully documented in Liberty Reservoir, it clearly is evident in Loch Raven and Prettyboy Reservoirs; see Interstate Commission on the Potomac River Basin, 2006).

In the case of Liberty Reservoir, the aforementioned seasonal pattern in DO concentrations follows a longitudinal gradient in the reservoir. The largest range of variation in the seasonal cycle of mean monthly DO concentrations appears to occur in the upper part of the reservoir, and the smallest range in variation in the lower part of the reservoir (Winfield and Sakai, 2003). Given the above, and that the range in annual variation chiefly relates to the extent of DO depletion, which reaches its maximum in the late summer to early fall, the frequency of severe DO biotic impairments (eutrophic conditions indicated by a TSI-DO index) in both shallow and deep layers would be expected to occur with the greatest frequency in the upper part of the reservoir, and with the least frequency in the lower part of the reservoir. This longitudinal pattern in severe DO impairments is not evident if TSI-DO index data are aggregated for the entire surface- or deep-water layers across the entire reservoir (table 8).

Oxygen impairment differs markedly between Prettyboy Reservoir or Loch Raven Reservoir and Liberty Reservoir in spatial extent, frequency of occurrence, and, to some extent, timing. A study by the Interstate Commission on the Potomac River Basin (2006) showed spatial differences in the degree of DO impairment in Prettyboy Reservoir occur primarily in late summer and early fall, with impairments most often occurring only in the deep-water layer in the lower and middle parts of the reservoir Prettyboy Reservoir (fig. 17). (Given no DO data have ever been collected in the upper part of this reservoir, it cannot be determined if these DO impairments occur in deep water throughout the reservoir.) In the case of Loch Raven Reservoir, annual variations in monthly mean DO concentrations for the period of record (Maryland Department of the Environment, 2006, **Appendix A**) indicate impairments also occur in the late summer to early fall, and routinely (almost every year) in the deep waters in the lower, mid, and upper parts of the reservoir, as well as in the shallow waters in most years, but slightly more often in the lower than middle or upper parts of the reservoir.

Relative to their evaluation of the monitoring data for DO concentrations for Prettyboy and Loch Raven Reservoirs, the Interstate Commission on the Potomac River Basin (2006) also noted that the use of a fixed-depth definition of 20–30 ft for shallow surface-layer waters (for example, table 8) can result in an inaccurate representation of DO concentrations in the true epilimnion, particularly in the fall (September through November). They found that water temperatures are an extremely sensitive indicator of stratification in these reservoirs, and modeled temperature profiles indicate that the well-mixed surface layer (epilimnion) is often less than 20 ft thick, and not up to 30 ft in depth. They also noted that the DO concentration for the surficial layer is lower if the greater depth interval is used to represent the epilimnion. This bias

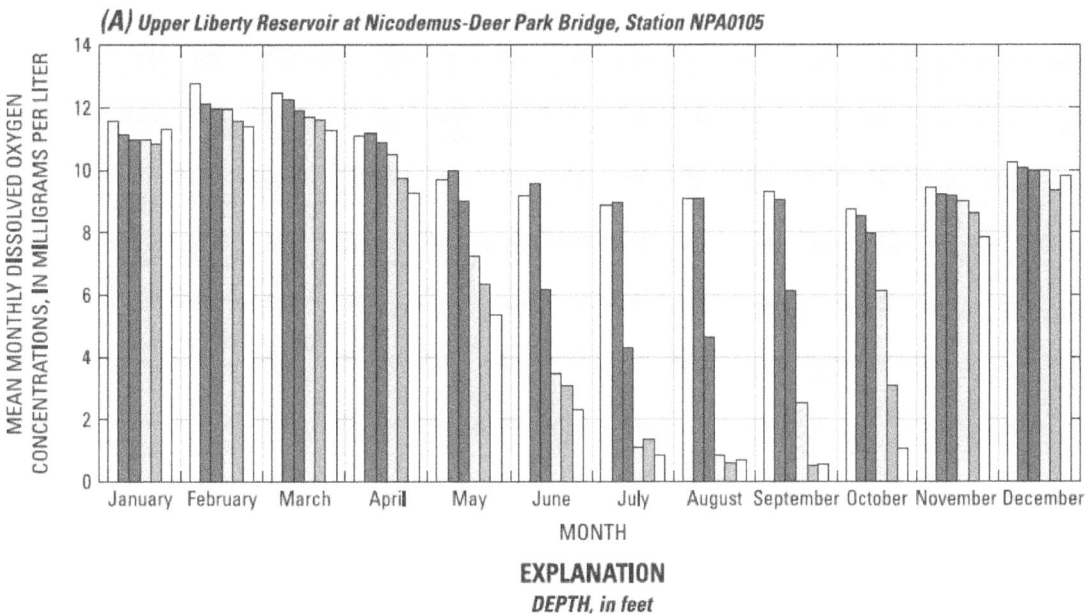

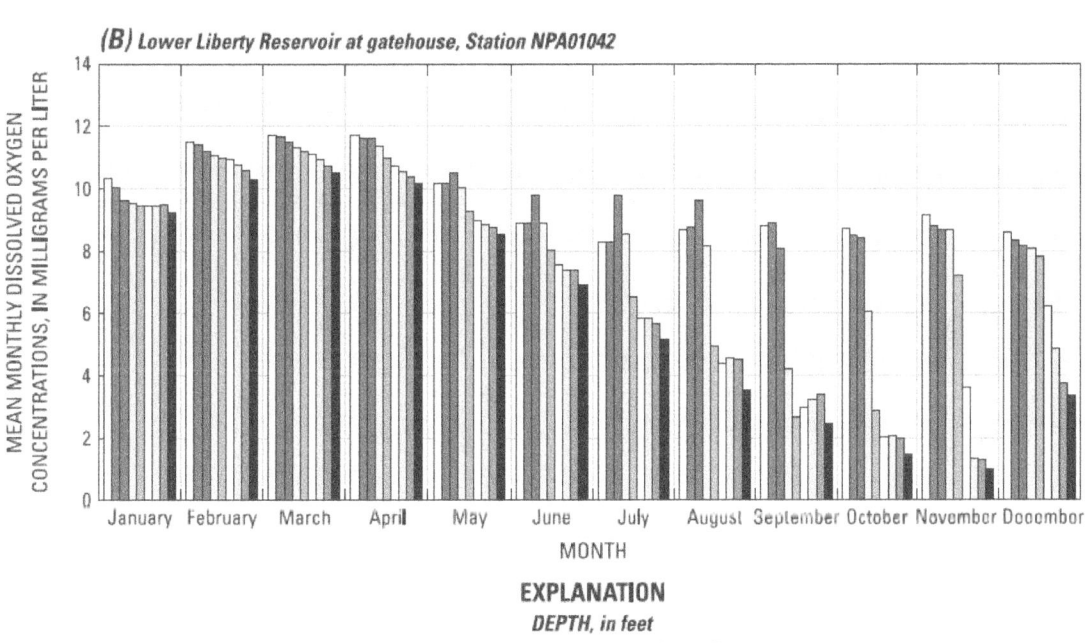

Figure 16. Mean monthly dissolved-oxygen depth-profile concentrations for *(A)* upper and *(B)* lower Liberty Reservoir, 1994–2001 (modified from Winfield and Sakai, 2003).

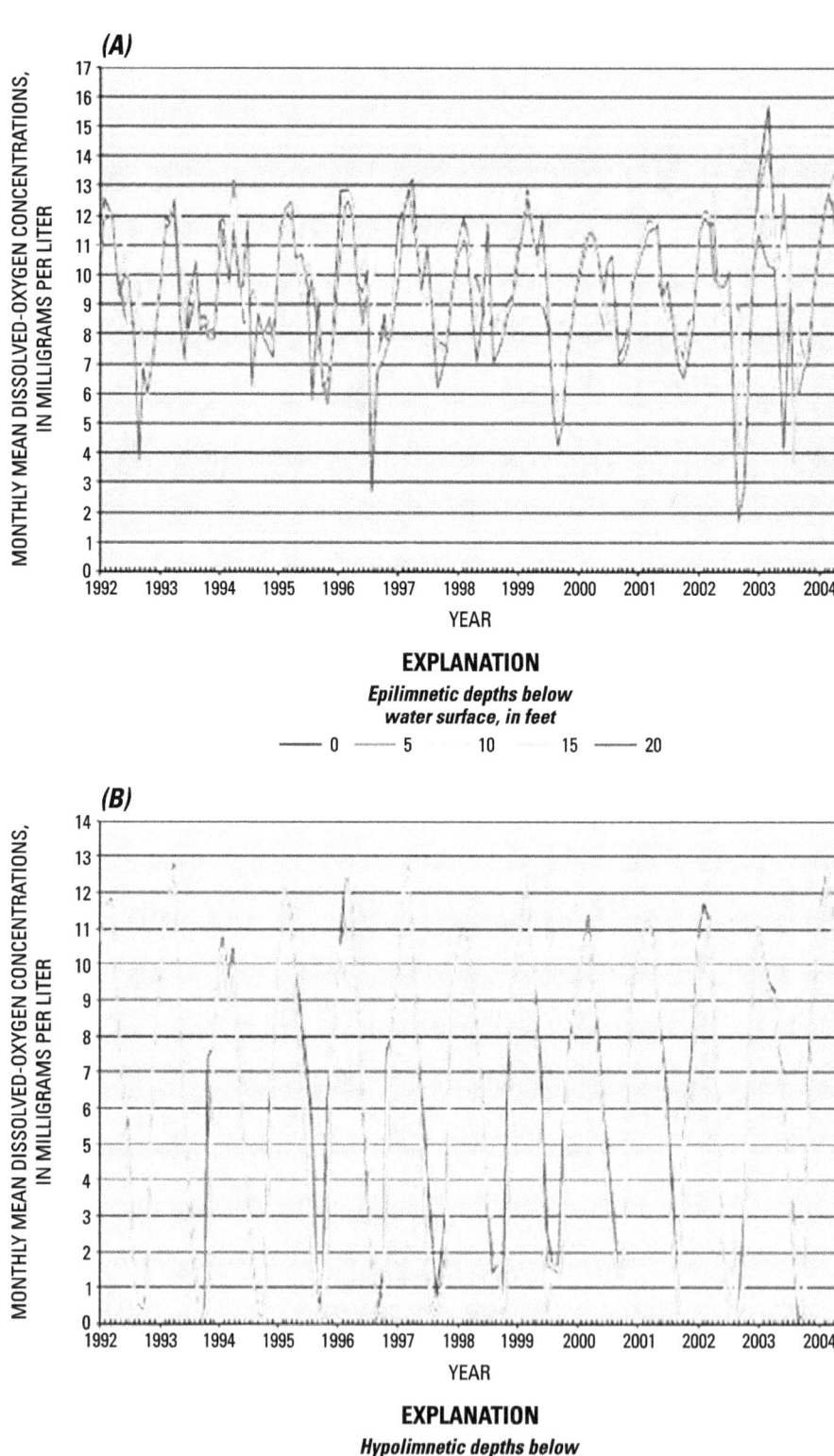

Figure 17. Monthly mean dissolved-oxygen concentrations in *(A)* epilimnetic waters and *(B)* hypolimnetic waters in lower Prettyboy Reservoir, 1992–2004 (modified from Interstate Commission on the Potomac River Basin, 2006). [Epilimnetic and hypolimnetic waters are defined by measured and modeled temperature gradients and simulated reservoir stratification and turnover.]

is particularly apparent during the fall because of episodic autumnal cooling. The cooling appears to result in temporary stabilization of the thermocline, which results in the entrainment of low-DO (metalimnetic) water into waters 20–30 ft below the lake surface, and thus the surface layer as traditionally defined by the City. Furthermore, in the case of Prettyboy Reservoir, the Interstate Commission on the Potomac River Basin (2006) indicated that similar problems appeared during the severe drought of 2002–03 (fig. 17) as a result of water releases from the reservoir.

As part of the effort to model DO concentrations in these two reservoirs, the Interstate Commission on the Potomac River Basin (2006) showed that the annual variation in DO concentrations, and the seasonal sags in DO, could be conditions indicative of internal reservoir hydrodynamics and therefore unlikely to change. This conclusion has important implications that go beyond the use of DO concentration data to determine trophic state to assess reservoir eutrophication and ecological health.

If seasonal DO sags are to be the norm, with deep-water DO concentrations falling below 1.0 mg/L, then addressing several questions raised earlier in this report takes on added significance: specifically, whether deep-water oxygen depletion results in anoxic conditions and iron, manganese, and phosphorus releases from bed sediments, and whether that released phosphorus routinely becomes available for future phytoplankton production. If these processes do occur as a result of natural hydrodynamic conditions, it could take a considerable reduction in TP tributary loads to reduce phytoplankton production as reservoir sediment contains an appreciable amount of bound phosphorus (see **Eutrophication— Nutrients: Phosphorus Transport and Reservoir Recycling**, and **Sedimentation—Reservoir Storage Capacity**, *this report*).

The TSI-DO index data to characterize the frequency of eutrophic and mesotrophic conditions (table 8) are likely biased—positively by only daytime DO measurements, and negatively by the manner in which DO data are aggregated and analyzed. The analysis of DO concentration data reveals temporal trends, seasonal variations, and spatial differences in DO, which likely influence the TSI-DO frequency of eutrophic conditions within each reservoir and among the reservoirs. None of the latter are apparent, however, given the manner in which the TSI-DO index data have been analyzed.

Water Transparency

Water transparency (as determined by Secchi disc depth to extinction) is used as an independent TSI indicator of ecological health in the Baltimore reservoirs (table 8). The Secchi-depth value corresponding to the Carlson TSI boundary between mesotrophic and eutrophic conditions is 6.1 ft. Application of this threshold to reservoir waters through late spring, summer, and early fall from the 1980s through the 1990s revealed each of the three reservoirs seldom exhibited a reduction in transparency that would correspond to eutrophic

conditions (table 8). In relation to other TSI indices, the Baltimore Reservoir Technical Group (2004) noted that transparency is the least sensitive indicator of TSI impairment during the spring-summer growing season (table 8).

Among the three reservoirs, the Baltimore Reservoir Technical Group (2004) and the Baltimore City Department of Public Works (2000) noted that water transparency appeared to decline in Prettyboy Reservoir during the 1980s through 1990s, but that there were no apparent trends in transparency in either of the other two reservoirs for the same data period. They postulated that the apparent decline in transparency in Prettyboy Reservoir could be the result of increased suspended sediment. The lack of stormflow monitoring data for tributaries to Prettyboy Reservoir, however, makes it difficult to test this hypothesis. They also did not address why this reservoir would not have a positive trend in transparency, or why no positive trends were observed in transparency in either supply reservoir, given that all three reservoirs appear to have had declining trends in algal counts and (or) chl-*a* concentrations throughout the 1990s (see **Eutrophication**, *this report*).

Seasonal patterns in transparency have been observed in all three reservoirs. The Interstate Commission on the Potomac River Basin (2006) described seasonal transparency patterns in Loch Raven and Prettyboy Reservoirs. Transparency generally was highest in the summer (typically approaching about 18 ft in depth in June and July), and then declined (by about two thirds) with lake turnover to its lowest levels in the winter or early spring (typically about 8 ft in depth or less from November through March) in both reservoirs. A similar annual pattern was observed in Liberty Reservoir (Winfield and Sakai, 2003), with the most pronounced seasonal differences in transparency occurring in the upper part of the reservoir. This result is consistent with seasonal and longitudinal patterns for chl-*a* concentrations (and algal counts), and for turbidity (suspended sediment), and likely reflects the fact that the major tributaries all discharge into the upper part of this reservoir.

A significant finding by the Interstate Commission on the Potomac River Basin (2006) in the transparency results for all three reservoirs is that light penetration likely only limits phytoplankton production throughout the epilimnetic layer (typically 20 ft or less) in the early spring. This result would explain why the TSI transparency index is not a very sensitive indicator of eutrophic conditions associated with major algal blooms in this layer compared to other TSI indices.

It also should be noted that the Interstate Commission on the Potomac River Basin (2006) studies of water transparency in Loch Raven and Prettyboy Reservoirs indicate there is an inconsistency between trends in transparency and trends in other water-quality indicators that should strongly correlate with transparency in the reservoirs. Transparency ought to be determined by turbidity caused by algal growth and reservoir turnover in combination with sediment transport by tributary watersheds. Thus, measurements such as chl-*a*, color, and turbidity should each be directly correlated with, and collectively explain most of, the variability in transparency. The

Interstate Commission on the Potomac River Basin (2006) essentially tested this hypothesis. Using an epilimnetic layer of 20 ft and mean monthly data for the above parameters from 1992–2004, they found that color or chl-*a* (algal source) and then turbidity (from non-algal sources, for example, suspended sediment) most likely account for reductions in transparency in Loch Raven Reservoir, and that turbidity, and then chl-*a* or color, most likely account for reductions in transparency in Prettyboy Reservoir. They also found that measurements of color, chl-*a*, and turbidity each accounted for no more than about 45 percent, and, collectively, accounted for only about 54 percent, of the total variability in transparency, in Loch Raven and Prettyboy Reservoirs. Thus, about half of the variability in water transparency in these two reservoirs cannot be explained.

One likely source of the unexplained variability in transparency could be differences in the measurement methods. Transparency data reflect a single depth-integrated monthly or bimonthly measure of the depth at which the Secchi disk disappears from view to an observer at each reservoir monitoring site. It is in effect a single continuous depth-integrated measurement, and, to a degree, a subjective (operator-sensitive) measurement. Measurements of chl-*a*, color, and turbidity are determined at a similar frequency, but each is an average value derived from discrete samples and instrumental measurements obtained at possibly three (near-surface, 10-ft, and 20-ft depth) discrete depths in the epilimnion at each reservoir monitoring site.

Because half of the variation in transparency in Loch Raven and Prettyboy Reservoirs cannot be explained by individual or collective measurements of turbidity, chl-*a*, and color, further investigation of measurement methods for all four parameters could be warranted. Studies of lakes or impoundments in North Carolina, Florida, and New Hampshire have found relatively high (individual or multivariate) correlation coefficients (ranging from approximately 50–90 percent) between transparency and one or more of these parameters (Caffrey and others, 2007; Schloss, 2002; Weiss and Kuenzler, 1976). For example, it could be that transparency measurements are not consistently obtained to visible extinction, or are obtained in a manner that introduces considerable variations in this measurement.

Sedimentation

The occurrence of elevated concentrations of suspended sediment in Baltimore Reservoir watershed tributaries and their contribution to sediment loads to the reservoirs are concerns for the RWMA partners (fig. 2). Elevated concentrations of suspended sediment in the tributaries reflect excessive losses of soil due not only to inadequately managed agricultural and urban lands, but also to the erosion of streambeds and banks mainly as a result of inadequately managed stormwater flows (**Appendix A**). The resultant damages to stream habitat and ecosystem health are considered impairments to designated recreational use of the tributaries, such as fishing (see **Regulatory Concerns**, *this report*).

The subsequent transport of tributary sediment to the reservoirs can lead to the loss of storage capacity—a designated use for the supply reservoirs. Sediment transported into each reservoir also can increase in-lake turbidity, which affects the cost of water treatment, and thus impairs the designated use of supply reservoirs for drinking-water supplies. Sediment deposited in the supply reservoirs also is a major potential source of iron and manganese (Ortt and others, 2000), metals, that when mobilized (see **Sedimentation—Sediment Diagenesis and Mobilization of Metals and Phosphorus**, *this report*), can increase the costs to treat water, as well as raise color and taste issues with treated water, which are considered impairments in the quality of reservoir water used for supplies (see **Regulatory Concerns**, *this report*). Recently deposited sediment, particularly during reservoir recovery following droughts, combined with the submergence of vegetation established on reservoir beds during droughts, also provides a readily available source of mercury for methylation and biological uptake. The mobilization, methylation, and uptake of this mercury could increase mercury concentrations in higher trophic order game fish. Although elevated concentrations of mercury in game fish is already a known impairment of designated recreational uses for all three reservoirs (see **Regulatory Concerns**, *this report*), this pathway of biological uptake has not been considered nor adequately investigated by the RTG. Sediment deposited in the reservoirs also contains elevated concentrations of phosphorus (Ortt and others, 2000), which if mobilized and available to phytoplankton, can help promote eutrophic conditions and excessive algal blooms in the reservoirs, which can cause additional impairments in designated recreational and water-supply uses for the reservoirs (see **Eutrophication**, *this report*).

To address most sedimentation-related concerns minimally requires quantifying sediment transport, identifying the main sources of sediment, and reducing sediment from these major sources. Thus, monitoring to describe sediment transport in the reservoir watersheds and to assess reservoir storage capacity is of critical importance to the City, RTG, and RWMA partners.

Sediment Transport

Development of the long-term watershed tributary monitoring network for suspended-sediment transport followed from the initial work of Amatayakul and others (1978), and Amatayakul, Defries, and others (1978), who provided some of the first estimates of total average annual loads for suspended sediment to Loch Raven Reservoir (table 4). In so doing, they noted a paucity of data for storms. They also cited a host of assumptions, other data limitations, and the need to rely on empirical data and methods from studies conducted elsewhere, to estimate loads. To address the above, they recommended a long-term monitoring program be designed and implemented. To accurately estimate loads and trends in loads, this program was to provide year-round collection of suspended-sediment samples for storm- and base flows on the major tributaries supplying the reservoir.

The recommendations of Amatayakul and others (1978) and Amatayakul, Defries, and others (1978) helped guide development of the long-term reservoir-watershed tributary monitoring network established in the early 1980s (fig. 3, and **Appendix B**). The resulting tributary monitoring network includes six gaged watershed water-quality monitoring stations for storm- and dry-weather flows, and nine additional dry-weather-flow tributary stations. The monitoring network also is used for sampling at selected point-discharge sites, including WWTPs, and at containment ponds if discharging, during routinely scheduled visits (**Appendix B**). In relation to the major subbasins that drain into each reservoir (fig. 3), the network includes water-quality monitoring for storm- and dry-weather flows at three large subbasins in each supply reservoir. Additional dry-weather-flow monitoring stations are used for selected small subbasins in the two supply reservoirs, and for the three major subbasins that drain into Prettyboy Reservoir. No monitoring occurs in small subbasins that are in close proximity and independently drain into each reservoir. The quality of water in most of these small subbasins occasionally or periodically has been monitored as part of short-term or synoptic studies conducted at different times by the State, or Baltimore or Carroll Counties (**Appendix A**).

The long-term monitoring of suspended sediment in the watershed tributaries has been useful in providing data and information on suspended-sediment concentrations, loads, trends, and potential source areas of sediment (among major subbasins). Tributary monitoring from the early 1980s to mid-1990s yielded data to assess annual variations in suspended-sediment concentrations and loads at each of the three major subbasins in each supply-reservoir watershed (figs. 18a and b). Summary reports indicate that the annual mean concentrations and total annual loads of suspended sediment markedly vary for each subbasin (Baltimore City Department of Public Works, 1992, 1996). This variability was attributed to climatic variation—concentrations and loads being notably higher in wet years compared to dry years (figs. 18a and b, for example, 1984 and 1986, respectively), and within a given year, to flow regime—notably more sediment being carried by stormflows as opposed to dry-weather flows, with the latter accounting for only 10 percent or less of the total annual sediment load (fig. 18b).

Suspended-sediment loads derived for the three major subbasins in each supply reservoir watershed during the 1980s also enabled the first spatial comparisons of loads. On the basis of areal-weighted loads (yields), subbasins within each supply reservoir were ranked to help identify major sediment-source areas and initially guide management to reduce sediment transport [Baltimore City Department of Public Works, 1996 (table 10.1), 2000, 2001]. For example, for the Loch Raven Reservoir watershed, it was noted that yields at the Gunpowder Falls at Glencoe station indicate that Prettyboy Reservoir captures most of the suspended sediment upstream from this station, and thus reduces the sediment load delivered to Loch Raven Reservoir. Therefore, management activities to reduce sediment loads to Loch Raven Reservoir could be

directed towards the main sources of sediment—the Beaver Dam and Western Run subbasins, and the lower Gunpowder River Basin between Prettyboy and Loch Raven Reservoirs (Maryland Department of the Environment, 2003d; Reservoir Watershed Protection Committee, 2000).

Monitoring data have been used to assess apparent trends in suspended-sediment concentrations and loads for the gaged tributaries. Suspended-sediment concentrations, and notably loads, appeared to decline after 1987 (figs. 18a and b). A visual comparison of suspended-sediment concentration-discharge curves constructed from monitoring data from 1983–85 and from 1986–90 for each of six major subbasins indicated less sediment was being transported throughout most of the range in flows after 1985 (Baltimore City Department of Public Works, 1992). No statistical tests accompanied the visual comparisons, however, to determine if the curves were significantly different.

The apparent decrease in sediment concentrations and loads, and downward shift in sediment rating curves, for both water-supply reservoirs were considered possible outcomes of a decade of best management practices (BMPs) that in part were designed and implemented to reduce erosion and improve agricultural and urban sediment control (Baltimore City Department of Public Works, 1992, 1996, 2000). It also was noted, however, that they could possibly reflect a bias due to the manner in which suspended-sediment data for storms were collected (Baltimore City Department of Public Works, 1996). No subsequent reports were encountered that indicated the nature of this change in methods, however, nor whether or not a change in monitoring methods was the cause or contributed to the decline in suspended-sediment concentrations, yields, or rating curves. Also, the alternative to tracking sediment-load reductions by tracking BMP installations and determining their efficiency (in the absence of relevant pre- and post-stream monitoring) proved difficult, and this accounting approach to sediment-load reductions resulting from BMPs was abandoned (Baltimore City Department of Public Works, 1996). Therefore, it was not possible to independently determine if the aforementioned decline in the concentrations, loads, and rating curves for suspended sediment truly resulted from the implementation of BMPs.

As noted earlier in this report, from about 1995 to 2004, monitoring of storm events was markedly reduced (table 7). The decline in stormflow data leaves a discontinuity in long-term records of total annual suspended-sediment loads from the three major subbasins in each supply reservoir watershed. Thus, advanced models to estimate suspended-sediment loads have few events for calibration or validation during most of this period—none in a number of years—despite almost an additional decade of BMP implementation to reduce sediment loads. For example, direct and local assessment of BMPs to demonstrate their effectiveness in reducing sediment transport, such as the recent study by Stewart and others (2005) on stream buffer restoration, strengthen the argument for the continued widespread monitoring and modeling of suspended-sediment concentrations to estimate loads.

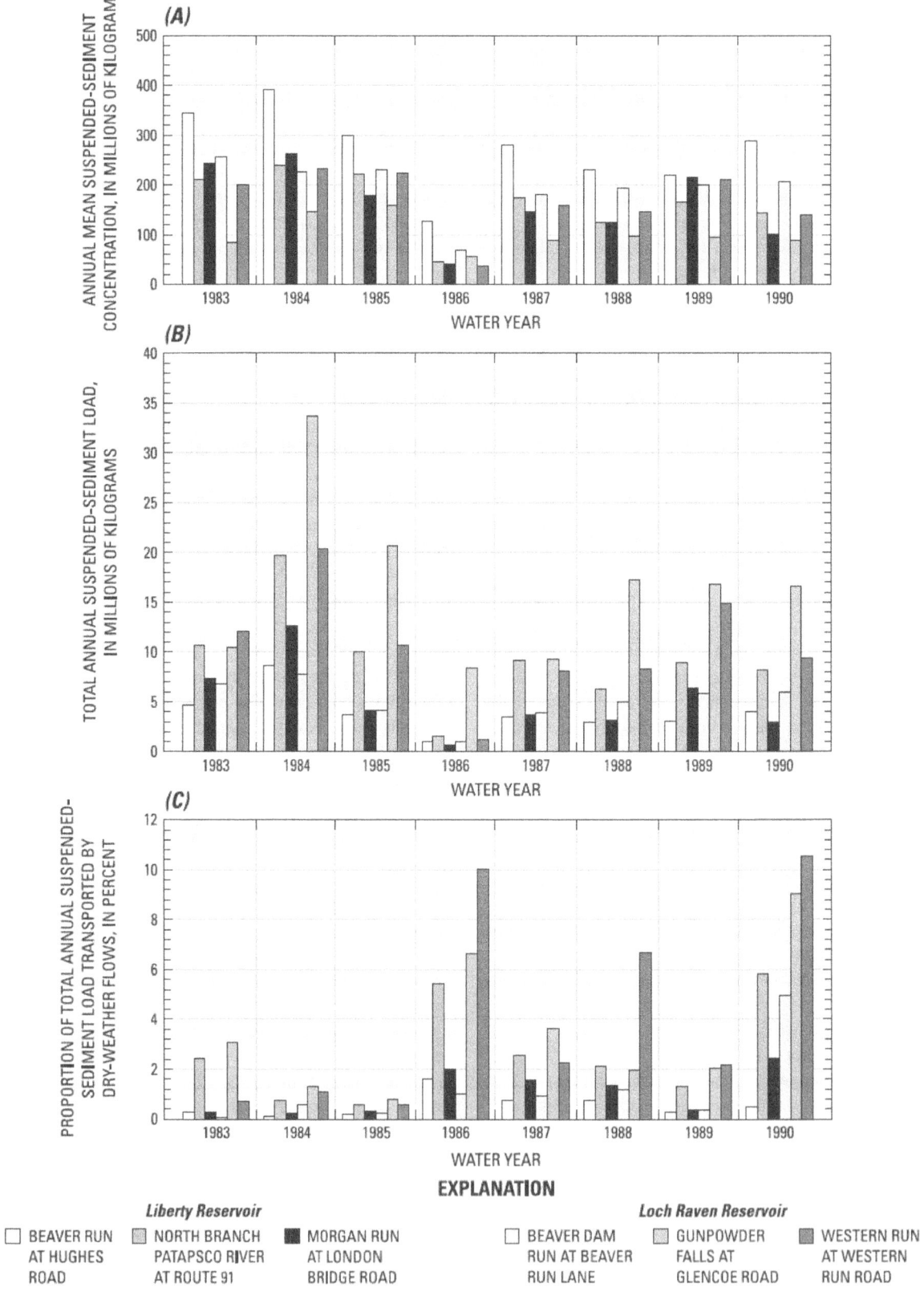

Figure 18.   *(A)* Annual mean concentrations of suspended sediment, *(B)* total annual suspended-sediment loads and *(C)* proportion of total annual loads transported by dry-weather flows for six gaged tributaries in subbasins in the Loch Raven and Liberty Reservoir watersheds, 1983–90 (modified from Reservoir Watershed Protection Subcommittee, 1992, and Baltimore City Department of Public Works, 1996).

A reasonably accurate and complete sediment record on major reservoir tributaries to properly calibrate and validate load-estimation models is vital to the RWMA partners. Modeled total annual suspended-sediment loads to Loch Raven Reservoir based on calibration periods of 1988–91 (Maryland Department of the Environment, 2003d) and 1992–97 (Interstate Commission on the Potomac River Basin, 2006) are approximately a thousand times greater than the initial estimates based on 1975–78 data from Amatayakul and others (1978) and Amatayakul, Defries, and others (1978) (table 4). In part, the differences in the estimated loads could reflect differences in the periods of record used for comparisons. Improvements in hydrologic and suspended-sediment records, combined with improvements in modeling, however, also likely resulted in more accurate load estimates that were substantially higher than earlier and less accurate load estimates.

Although storm sampling has been limited since 1995 (table 7), dry-weather monitoring of suspended-sediment concentrations and flows, which also began in 1983, generally has continued almost uninterrupted at all stations in the three reservoir watersheds. Because dry-weather-flow data represent just part of the overall flow and sediment-transport regime, their utility is limited (Baltimore City Department of Public Works, 1992, 1996, 2000, 2001). For the three major subbasins in each supply reservoir watershed, the dry-weather flow data have enabled the determination of the proportion of the total annual sediment load actually transported by dry-weather flows (fig. 18b). These data also were used to derive suspended-sediment concentration-discharge curves for the six major subbasins, which were described above and indicated a reduction in sediment loads after 1986.

Suspended-sediment data from dry-weather flows also have been used to compare annual yields from subbasins in all three reservoir watersheds to assess apparent seasonal variations or annual trends among the subbasins, and whether these apparent variations or trends relate to trends in reservoir water quality (Baltimore City Department of Public Works, 1992, 1996, 2000, 2001). Inherent in the use of dry-weather-flow suspended-sediment data for the above purposes, however, is the assumption that they have and remain highly, positively, and similarly correlated with the corresponding yields, seasonal variations, and trends in suspended sediment associated with storm- and dry-weather flows at each dry-weather-flow monitoring station. No analyses were encountered in this retrospective review that validate these assumptions. The historical and continued use of dry-weather-flow suspended-sediment data for the above purposes requires that such correlations be demonstrated. Historically, it is already apparent that there has been at least one indication that the proportion of the total sediment load carried by dry-weather flows has changed (fig. 18c). If this change can be statistically verified, and it is due to BMPs rather than sample-collection methods, then additional changes in the proportion of suspended sediment

carried by dry-weather flows could be expected. Such a finding also would preclude the use of dry-weather flows for the above purposes.

Accurate estimates of seasonal variations in suspended-sediment concentrations, and computation and comparisons of annual trends in loads or yields, generally require the measurement of suspended-sediment concentrations during both storm- and dry-weather flows at a majority of, if not all, reservoir watershed subbasins. For example, Gellis and others (2005) indicated that comparisons of loads and yields among Chesapeake Bay watershed tributaries generally required each tributary have suspended-sediment records for at least 3 consecutive years that included approximately 10 storms per year.

Reservoir Storage Capacity

Summary interim and final reports from the City Department of Public Works (Baltimore City Department of Public Works, 1996, 2000, and 2001) and the Baltimore Reservoir Technical Group (2004) indicate that sedimentation in the reservoirs is a long-term and continuing concern. Recent bathymetric surveys (1998–2000) indicate all three reservoirs have lost some storage capacity since their initial construction—from about 3 percent to 11 percent depending on the reservoir (table 1). In at least the case of the Loch Raven Reservoir, sediment deposited by the year 2000 likely had filled available storage in the lower parts of the watershed tributaries and certain upland parts of the upper reservoir (Ortt and others, 2000).

One main implication and concern is that future sediment transport and deposition could occur further into the reservoir, and accelerate the loss in reservoir capacity. Therefore, the RTG plans to monitor the loss in the storage capacity of Loch Raven Reservoir and the other two reservoirs through future surveys conducted every 10–15 years (Baltimore Reservoir Technical Group, 2004). As part of their findings, and in regard to these future surveys, Ortt and others (2000) indicated that their surveys were hampered by submerged aquatic vegetation because they were conducted during the summer months. They also noted that they were unable to locate or utilize historical bathymetric survey data from several previous surveys because of the inadequate manner in which survey data were archived, and survey methods were documented.

Another implication and concern regarding sedimentation expressed by the Baltimore Reservoir Technical Group (2004) is that the dispersion of sediment further into the reservoir could increase the potential mobilization of metals including iron, manganese, and mercury, as well as nutrients such as phosphorus, which all occur in reservoir sediment (Ortt and others, 2000). On the basis of this review, the decreased storage is almost certain, but the possible increase in metal and phosphorus availability requires the RTG to address the larger question of the extent to which metal and phosphorus mobilization from sediment occurs in the reservoirs in general.

Sediment Diagenesis and Mobilization of Metals and Phosphorus

How available are the metals and phosphorus associated with sediment transported and deposited in the reservoirs? The answer is not clear. Ortt and others (2000) established that reservoir sediments contain elevated concentrations of metals such as iron and manganese and elevated concentrations of phosphorus, which appears mainly bound to iron in these sediments.

Except for manganese and iron in reservoir intake waters, and possibly mercury in game fish, summary and other reports (Baltimore City Department of Public Works, 1996, 2001; Maryland Department of the Environment, 2003 a,b,c; Baltimore Reservoir Technical Group, 2004) indicate that heavy metals are not a concern in reservoir waters (see **Regulatory Concerns**, *this report*). Elevated concentrations of iron and manganese in reservoir intake waters can lead to nuisance problems in the treatment of water, and odor, taste, and appearance issues in finished water. Their origin and pattern of occurrence as elevated concentrations in reservoir waters have been discussed (see **Overview of Water-Quality Concerns**, *this report*). Their primary source appears to be the release of iron and manganese as a result of sediment diagenesis under anoxic conditions. Anoxia occurs with deep-water DO depletion caused at least in part by the decomposition of algal residues following major blooms and die-offs (Winfield and Sakai, 2003). Thus, reductions in impairments to drinking water caused by iron and manganese largely will be controlled by the degree to which eutrophic conditions that lead to excessive algal blooms can be reduced. Reducing major algal blooms will help reduce deep-water DO depletion during bloom decomposition, and thus reduce the anoxic conditions that lead to the release of these metals from reservoir sediments.

Elevated concentrations of mercury have been found in upper-trophic-level game fish in all three reservoirs, and notably in largemouth bass (*Micropterus salmoides*) tissues. Bass fish-tissue concentrations in the reservoirs range from minimums between approximately 50–100 µg/kg (micrograms per kilogram) to maximums between approximately 400–800 µg/kg; the geometric mean concentrations for fish tissues in the three reservoirs ranges from approximately 260–310 µg/kg (Maryland Department of the Environment, 2002 a,b,c; 2004 b,c). Mercury concentrations in game fish tissue exceed the Maryland State (235 µg/kg) and Federal (300 µg/kg) CWA standards. Combined with an upward revision in risk assessment for mercury consumption, the City had to take action (Maryland Department of the Environment, 2002 a,b,c; 2004 b,c). In addressing this reservoir impairment, the City provided data that indicated the bulk (85–95 percent) of the mercury delivered to the reservoirs and watersheds is from atmospheric deposition, and argued that the source of this mercury is regional and largely beyond control. As of 2007, the City findings were under review by the USEPA.

Not considered in the arguments presented by the City but worthy of note is that during periods of low water levels in the reservoirs, such as those that occur following prolonged withdrawals during drought conditions, exposed reservoir bed sediments can be oxidized, and subject to extensive re-vegetation by terrestrial plants (fig. 19). The area of exposed and potentially vegetated and oxidized bed sediments in the City reservoirs during prolonged droughts is considerable given the extent of reservoir shorelines (table 1, fig. 3B). Submergence of this vegetation and oxidized sediments during reservoir recovery could lead to what has been described as a first-flush release of nutrients (Newman and Pietro, 2001) and trace metals, including mercury, from recently deposited reservoir bed sediments (Dmytriw and others, 1995; Rawlick, 2001). In the presence of freshly released and elevated concentrations of oxidized carbon, sulfur, iron, and bioavailable Hg (II), there can be an acceleration in methylmercury production (Morrel and others, 1998; Krabbenhoft and others, 2000; Krabbenhoft and Fink, 2001), leading to increased biological uptake. Whether this process is an internal contributing factor to the elevated concentrations of mercury found in game fish in the City reservoirs cannot be determined without adequate data being obtained before (baseline) and after (within first 2 years of) recovery conditions in those areas of the reservoir where terrestrial vegetation is inundated. In general, the occasional monitoring of mercury in game fish in the reservoirs is inadequate in terms of timing and spatial distribution to determine whether enhanced mercury methylation occurs in those shallow areas of the reservoirs subject to exposure, re-vegetation, and subsequent re-inundation during drought and recovery periods. It also is unlikely that the answer to this question can be inferred from other Maryland lakes or reservoirs. There is a markedly wide range in the concentrations of methylmercury found in game fish in most Maryland lakes and reservoirs that cannot be explained, and likely indicates highly variable rates in regional atmospheric deposition and (or) reservoir response (Sveinsdottir and Mason, 2005).

Mobilization of phosphorus from reservoir sediments has been of concern to the RTG because of its possible role as the limiting nutrient in phytoplankton production (see Eutrophication—**Nutrients: Phosphorus Transport and Reservoir Recycling**, *this report*). Four studies have discussed its potential occurrence; but collectively the results of these four studies are ambiguous in terms of whether phosphorus release occurs, and if it occurs, whether the released phosphorus becomes available for phytoplankton production. In the first of these studies, Amatayakul and others (1978) conducted an analysis of Loch Raven Reservoir water-column data from which they concluded that bottom sediments were not a likely source of the phosphorus for algal blooms. From the concentrations of TP observed at depth, they concluded that reservoir bottom sediments appeared to release phosphorus. But they also argued that the released phosphorus likely would be unavailable, because during the summer months, stratification would prevent its utilization in the epilimnion, where algal blooms most often occur. Furthermore, they theorized that during lake turnover in the fall and winter, this phosphorus could reach shallow depths, but the reduced temperatures and

April 26, 2002

April 26, 2002

August 4, 2002

August 4, 2002

Figure 19. Terrestrial vegetation growing on exposed in-lake bed sediments in upper Prettyboy Reservoir in 2002. [Photographs by Wendy S. McPherson and Michael T. Koterba, U.S. Geological Survey.]

day length would limit bloom occurrence. Their study, however, was limited in a number of ways. Most of their conclusions were drawn from only a few samples collected in 1 year (1973). The samples were analyzed for just TP, and not bioavailable phosphorus. The samples reflected selected depths at just three sites in just Loch Raven Reservoir. Although these samples were collected before and during reservoir stratification, examination of the data indicates that the concentrations of DO were measurable at all sampled depths in the hypolimnion. Hence, the anoxic conditions that were suspected to occur at depth in the reservoirs in association with the die-off of major algal blooms, and which are necessary for the release of phosphorus from reservoir sediments, actually were not truly investigated. Finally, in generalizing their findings, the authors did not know at the time that major algal blooms can occur in Prettyboy Reservoir during the late winter or early spring (Interstate Commission on the Potomac River Basin, 2006), which could reflect phosphorus released from bottom sediments and made available through lake turnover the preceding fall and winter.

In the second study, Winfield and Sakai (2003) noted that for a number of years from 1981–93, concentrations of TP in deep water (greater than 30 ft deep) exceeded concentrations of TP in shallow waters (30 ft deep or less) in Liberty Reservoir. They interpreted elevated TP concentrations at depth to be the result of phosphorus released from reservoir bottom sediments. They indicated that the typical spring blooms in this reservoir could be in response to nutrients possibly made available from lake turnover the preceding fall and winter.

In the third study, Ortt and others (2000) found that sediments in all three reservoirs contained abundant phosphorus, and that surficial bottom sediments appeared to be enriched in phosphorus. Sediment phosphorus concentrations were high, and highly correlated with those of iron, and, therefore, likely bound to iron. They did not postulate whether this phosphorus would become available for algal blooms.

In the fourth study (Interstate Commission on the Potomac River Basin, 2006), the investigators found that anoxic conditions (no measureable DO) did occur at depth during the late summer and early fall in both Loch Raven and Prettyboy Reservoirs (for example, see fig. 16B), indicating that iron-bound phosphorus could be released under these conditions. They noted that this would occur, however, only after oxic nitrogen species (for example, nitrate) are reduced (see also Hem, 1970; Wetzel, 2001). But they also pointed out that nitrate generally was always present at measurable (above 0.5 mg/L) concentrations at reservoir-bottom monitoring stations. They theorized that if nitrate is sufficiently abundant to serve as the preferred electron acceptor over ferric iron, iron would not be mobilized, and phosphorus would remain bound to iron in bottom sediments in the reservoir. They do not consider, however, whether the measurements taken at only a few locations in the deep parts of these reservoirs adequately reflect bottom conditions throughout the entire reservoir. Is nitrate present at all depths throughout each reservoir during periods of anoxia?

Collectively, these four studies appear to leave several unanswered but critical questions. What are the sources of available phosphorus that lead to algal blooms, and do they include the carryover of bioavailable phosphorus from one year to the next? For example, does lake turnover mobilize phosphorus from biomass decay or sediment release in one year to contribute to spring algal blooms in the following year, or is the source of the bioavailable phosphorus just winter and spring streamflows? Within a year, are mid-to-late summer algal blooms the result of bioavailable phosphorus released in part by decay of biomass from earlier algal blooms and (or) released from bed sediments in late summer? The answers to these questions are not known with any certainty because other than TP, no long-term routine measurements have been made in both the tributaries and reservoirs of bioavailable phosphorus, such as DOP or TDP.

Distribution and Sources of Bacteria and Other Potential Pathogens

Summary reports (Baltimore City Department of Public Works, 1996, 2001; Baltimore Reservoir Technical Group, 2004) indicate that although bacteria have been found at elevated counts in selected locations in the watershed tributaries, the Baltimore reservoirs appear to function in an effective manner to reduce microbial contaminants at supply intakes. Individual sample values and monthly averages for fecal coliform bacteria counts in raw water from Liberty and Loch Raven Reservoirs consistently fall below and well below,

respectively, the State water-quality recreational water-contact standard (200 Most Probable Number per milliliter). As of 2007, selected watershed tributaries in each reservoir watershed had bacterial concentrations that exceeded this standard, and MDE had requested additional data be collected and analyzed (see **Regulatory Concerns**, *this report*). Pending the outcome of this request, long-term monitoring in the reservoir watershed tributaries could require the collection of fecal-coliform or other bacterial data. The design of the bacterial monitoring program for the reservoir watershed tributaries (if it follows other similar multi-watershed-scale studies) could depend on the type and levels of bacteria found, their spatial and temporal distribution, and their potential sources (Francy and others, 2000). For the Baltimore Reservoir System watersheds, likely potential source areas include agricultural as well as (sub)urban lands, with potential sources being livestock and pet wastes (manure), aging septic systems (rural), sewer and outfall lines [(sub)urban, residential, commercial, and industrial], storm drains and lines [rural and (sub)urban)], and WWTP storm-related overflows (Valcik, 1975).

Emerging Water-Quality Concerns

Emerging water-quality concerns addressed in this review and evaluation and described in recent (post-2000) RWMAs include the occurrence of elevated concentrations of DBPs in treated water in the distribution systems associated with both water-supply reservoirs. Recent increases in sodium and chloride in reservoir and treated waters are another emerging concern, and presumably result from the increased use of sodium-chloride salt as a deicing agent. In addition, although not formally recognized in the RWMAs as an emerging concern, the effects of climate are considered in this review and evaluation as an emerging concern. The projected changes in climate combined with the inherent variability in climate in the region could produce marked fluctuations in tributary and reservoir water quality, and therefore need to be considered in relation to current and future monitoring.

Disinfection By-Products

Disinfection by-products (DBPs) are an emerging concern at the City water-treatment plants because of recent (2006) changes to the manner in which the USEPA SDWA standards governing the concentrations of these compounds must be applied to treated drinking water (see **Regulatory Concerns**, *this report*). As of 2007, little is known about the source(s) of organic carbon found in Baltimore reservoir intake waters, which form DBPs during or after the initial disinfection (chlorination) of raw intake water or the residual chlorination of water after coagulation and filtration, which is used to maintain the quality of water as it moves through the City water-distribution system.

DBPs typically form when organic carbon, both dissolved and particulate, reacts during or after chlorination to form chlorinated and brominated compounds, which have been shown to be harmful to human health. From an operational standpoint, total organic carbon (TOC) is employed by the City as a surrogate for the organic matter that reacts upon chlorination to form DBPs. A TOC concentration exceeding 2.0 mg/L in reservoir intake waters is the action level that requires treatment-plant operations be undertaken to reduce TOC, and thus possibly DBP formation.

Although there are a wide array of DBPs potentially formed during treatment-plant operations (Krasner and others, 2006), only two of the major categories of DBPs—trihalomethanes (THMs) and haloacetic acids (HAAs)—and only the total concentration of the most commonly observed compounds in each of these classes—are regulated by USEPA SDWA standards.

Because regulations regarding DBPs relate to total THM and total HAA concentrations found in drinking-water distribution systems, most of the THM and HAA data routinely obtained by the City DPW reflect finished-water samples collected after treatment, and within the water-supply distribution systems. Therefore, the reported DBP concentrations reflect variations in the quantity of carbon that forms DBPs in raw intake waters, and variations in treatment-plant operations. For the City treatment-plant operations, the latter typically can include the initial chlorination, coagulation, and filtration of water, as well as the possible temporary storage and possible re-chlorination of finished water, and ultimately the transport of the treated water through the water distribution system.

The occurrence of DBPs in the Baltimore reservoir water-supply system is an emerging concern because of the need to adopt changes in Federal regulations regarding the reporting of total THMs and total HAAs in public water-supply distribution systems. Under the existing Federal rule, routine monitoring by the City DPW for DBPs indicates that the 30-day moving-average total concentrations of THMs or HAAs have not exceeded their respective USEPA MCLs of 80 μg/L and 60 μg/L, respectively (Baltimore Reservoir Technical Group, 2004). The City DPW has noted, however, that summer-to-fall mean THM concentrations in the distribution system during the mid-1990s sometimes approached the 80 μg/L USEPA MCL for THMs, and that during 1995–2001, total THM and or HAA concentrations at selected stations occasionally exceeded their respective USEPA MCLs under what would be the revised reporting rules. Revised rules will require the MCLs be applied to the 30-day moving average of total THMs or total HHAs at each point of sampling within the water-distribution system.

The Baltimore Reservoir Technical Group (2004) indicated additional studies were needed to identify sources of DBP precursors. In support of these studies, Winfield and Sakai (2003) provided some initial DBP-related data on Liberty Reservoir intake waters at the Ashburton treatment facility (table 9). MDE (2004a) also summarized DBP data for the Loch Raven Reservoir Montebello treatment facility (table 10).

The data from Winfield and Sakai (2003) indicate that, except for 1998, water treatment at the Ashburton plant, which utilizes water from Liberty Reservoir, reduced the initial

concentration of TOC by approximately 30 percent, but that considerable TOC, approximately 1.7–2.9 mg/L, still remained in the water after coagulation and filtration (table 9). Also the concentrations of total THMs and total HAAs encountered in finished waters appear to be solely a result of the treatment process, as no THMs or HAAs were detected in raw intake waters.

Data on DBPs and TOC in finished waters analyzed by the Maryland Department of the Environment at the Montebello treatment facility (table 10) generally reflect water from Loch Raven Reservoir, or from the Susquehanna River when the reservoir water levels are low (see **Reservoir Watershed Management and Reservoir Operation**, *this report*). The DBP data from the Montebello treatment facility are consistent with data presented above for the Ashburton treatment facility. Overall MDE found that the average annual total concentration of THMs or HAAs from all samples collected at all stations were below the respective USEPA MCLs. Maximum annual concentrations, which reflect conditions at individual stations, exhibited total THM and HAA concentrations that would have exceeded the respective USEPA MCLs under the revised rules. MDE (2004a) also summarized data on a quarterly basis that indicate about 30 percent of the TOC in raw supply waters at the Montebello treatment facility was removed by coagulation and filtration.

To aid in understanding the occurrence of elevated DBPs in the City water-supply distribution system, the USGS conducted an analysis of recent (2003–08) data provided by the City DPW on THMs, HAAs, and TOC. The purpose of this retrospective analysis was twofold:

a) To describe and characterize the THM, HAA, and TOC data, in relation to concentrations and trends at each water-treatment facility, and

b) To identify in the broadest terms the potential sources (reservoir or watershed) of carbon that forms DBPs.

There are inherent limitations in the analyses of the City DBP and TOC data. Whereas TOC is routinely monitored at the reservoir intakes, there has been no routine long-term in-lake or tributary monitoring for TOC specifically focused on the identification of the carbon that forms DBPs in finished waters. Nor has the timing and frequency of routine sampling used by the City DPW specifically been designed to relate the TOC concentration found in an intake water to the total THM or HAA concentrations in this water after it has been treated and distributed. Thus, although sampling for TOC and DBPs could have been conducted on the same day, the intake water that was collected and analyzed for TOC is not necessarily the finished water that was sampled for DBPs.

Also relative to the THM, HAA, and TOC data that have been collected, much of the detailed ancillary information on variations in actual plant operations that could influence DBP formation is unavailable. In the case of the Baltimore water-supply system and treatment-plant operations these variations include:

Table 9.   Annual maximum concentrations of total organic carbon, trihalomethanes, and haloacetic acids in raw and treated waters in the Ashburton treatment plant, Liberty Reservoir, 1997–2001 (modified from Winfield and Sakai, 2003).

(TOC, total organic carbon; THMs, trihalomethanes; HAAs, haloacetic acids; mg/L, milligrams per liter; µg/L, micrograms per liter; ND, not detected)

Year	Annual maximum concentrations					
	TOC, raw water (mg/L)	TOC, treated water (mg/L)	THMs[1], raw water (µg/L)	THMs, treated water (µg/L)	HAAs[2], raw water (µg/L)	HAAs, treated water (µg/L)
1997	2.54	1.69	ND	40.0	ND	35.0
1998	2.00	2.70	ND	27.0	ND	46.0
1999	2.62	1.38	ND	36.0	ND	38.0
2000	3.42	2.30	ND	36.0	ND	33.0
2001	3.10	2.92	ND	19.5	ND	34.0

[1] Sum of the concentrations of chloroform, bromodichloromethane, dibromochloromethane, and bromoform.

[2] Sum of the concentrations of mono-, di-, and tri-chloroacetic acids and mono- and di-bromo acetic acids.

Table 10.   Summary statistics for annual concentrations of regulated total trihalomethanes and haloacetic acids in finished waters at the Montebello treatment facility, Loch Raven Reservoir, 1996–2003 (modified from Maryland Department of the Environment, 2004a).

[All concentrations are in micrograms per liter; ---, unavailable]

Year	Concentration of trihalomethanes[1]				Concentration of haloacetic acids[2]			
	Average	Maximum	Minimum	Number of samples	Average	Maximum	Minimum	Number of samples
1996	45	82	20	96	---	---	---	---
1997	40	84	17	45	---	---	---	---
1998	40	77	15	54	33	58	16	35
1999	47	93	15	36	19	41	2	27
2000	41	80	23	36	48	102	1	28
2001	50	100	22	18	28	53	7	16
2002	52	87	17	42	26	58	0	42
2003	33	54	21	11	40	59	13	11
Total	**44**	**100**	**15**	**338**	**31**	**102**	**0**	**159**

[1] Sum of the concentrations of chloroform, bromodichloromethane, dibromochloromethane, and bromoform.

[2] Sum of the concentrations of mono-, di-, and tri- chloroacetic acids and mono- and di- bromo acetic acids.

a) Variation in the sources of raw water, which in the case of the Montebello treatment facility, generally is from Loch Raven Reservoir but occasionally, and particularly during droughts, could include water from the Susquehanna River;

b) Variation in treatment plant operations, including the levels of chlorination and possibly re-chlorination, given the latter is sometimes required;

c) Variation in the residence time of stored treated water, which for each treatment facility can be held in outdoor finished-water ponds for up to several days, before it is possibly re-chlorinated, and then placed into the distribution system;

d) Variation in the travel (residence) times of the treated water after it enters the distribution system; and

e) Variation in the formation and degradation rates of THMs or HAAs.

Since there is no permanent documentation on the long-term history of these variations at each treatment plant, there also is no way to rank them according to their relative importance to DBP formation. It generally is known, however, that variations in each of the above factors can affect the concentrations of total THMs or total HAAs found in finished water within different parts of a large water-supply distribution (storage and pipeline) system (Singer and others, 2002).

Given these limitations, the monitoring data for DBPs for individual stations and TOC in raw intake waters only can be summarized, described, and analyzed in a broad manner. For the period of record (January 2003–July 2008), the total concentrations of THMs and HAAs in finished-water samples periodically exceeded the USEPA MCLs at one or more stations in each water-supply distribution system (figs. 20 and 21). The total concentration of THMs at one or more stations exceeded 80 µg/L on approximately 19 percent of the sample-collection dates for either distribution system; the total concentration of HAAs exceeded 60 µg/L at one or more stations on approximately 43 percent and 38 percent of the sample-collection dates for the Ashburton and Montebello distribution systems, respectively (table 11). In addition, for most sampling dates, about the same number of stations in both distribution systems were likely to have total THM (fig. 20) or HAA (fig. 21) concentrations that exceeded their respective USEPA MCLs on that date.

The concentrations of THMs in finished waters associated with either treatment plant appear cyclic in nature on a seasonal basis (fig. 20). For the period of record, the highest THM concentrations, and thus frequency of exceedances of the USEPA THM MCL, generally occurred each year from mid-summer through early fall (July–September). MDE (2004a) found a similar cyclic pattern for THM exceedances for 1995–2003 DBP data collected from the Montebello distribution system.

The repeated frequencies with which the total THM concentration exceeds the USEPA MCL for THMs at the same stations during mid-summer through early fall indicate that it is during this period that a 30-day moving average THM concentration at these stations is most likely to exceed the USEPA for THMs under the pending rule change. In contrast, seasonal trends in the total concentration of HAAs are not readily apparent (fig. 21). Elevated total concentrations of HAAs often occur at the same stations, but at different times in different years. Thus, it is uncertain whether there will be no periods, or occasional but random periods, during a given year or from year to year during which a 30-day moving average HAA concentration at each of these stations is likely to exceed the USEPA MCL for HAAs under the pending rule change.

Interpretation of the likelihood that the 30-day moving average for the total concentration of either THMs or HAAs will exceed their respective USEPA MCL is further complicated by the fact that in either distribution system, the total concentrations of THMs and HAAs during the period of record (January 2003–July 2008) at a given station are largely independent of one another. There is little to no significant linear correlation between the total concentration of THMs and the total concentration of HAAs at a given distribution station. For 19 stations, linear correlation coefficients (R^2) for most (17/19) stations were less than or equal to 0.10, with p values equal to or greater than 0.10; for two stations, the R^2 values were between 0.30–0.40, and p values were less than 0.002, but linearity was largely the result of one or two outlier values (linear regression analysis, Helsel and Hirsch, 2002). This result is not atypical given that THMs generally form and degrade more quickly than HAAs in large distribution systems (Xie, 2004).

As for the sources of carbon that form DBPs, concentrations of TOC are the only routinely measured form of carbon that can serve as a potential indicator of the occurrence of DBP precursors in the City water-supply distribution systems. Excluding an extreme outlier value of 9.6 mg/L recorded on October 13, 2005 at the Montebello treatment facility, concentrations of TOC at the two treatment facilities generally were similar. For the period of record, at the Montebello facility, they ranged from 1.3 to 3.5 mg/L and averaged 1.9 ± 0.3 mg/L; at the Ashburton facility, they ranged from 1.2 to 3.0 mg/L and averaged 2.0 ± 0.3 mg/L.

For the period of record, variations in the concentration of TOC in treated water were generally similar to variations in the concentration of TOC in raw-intake waters at either treatment facility (fig. 22). The main difference in these two measurements of TOC is that treated-water TOC concentrations were consistently lower than raw-water TOC concentrations. The treatment process, which typically involves chlorination, coagulation, and then filtration, routinely removed about one-third of the total carbon found in raw-intake waters. This result is similar to results described earlier by Winfield and Sakai (2003) for the Ashburton treatment facility from 1997–2001, and by MDE (2004a) for the Montebello treatment facility for TOC data collected from 1996–2003.

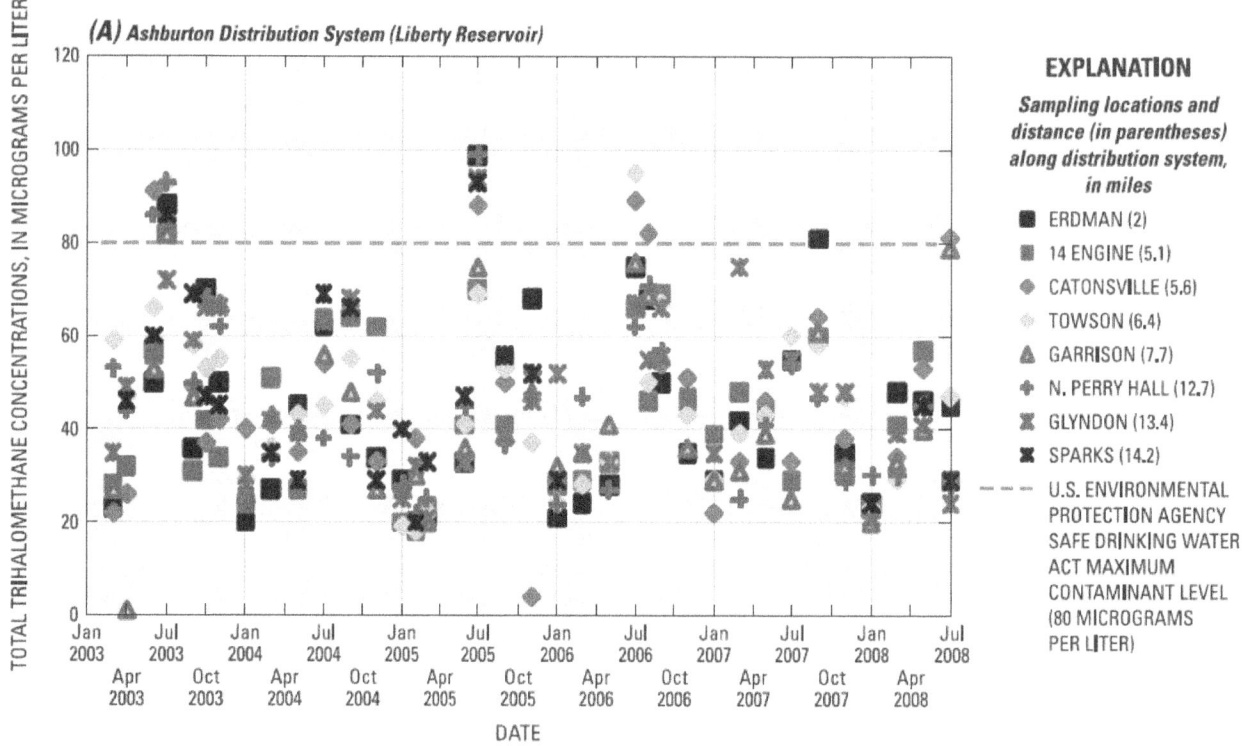

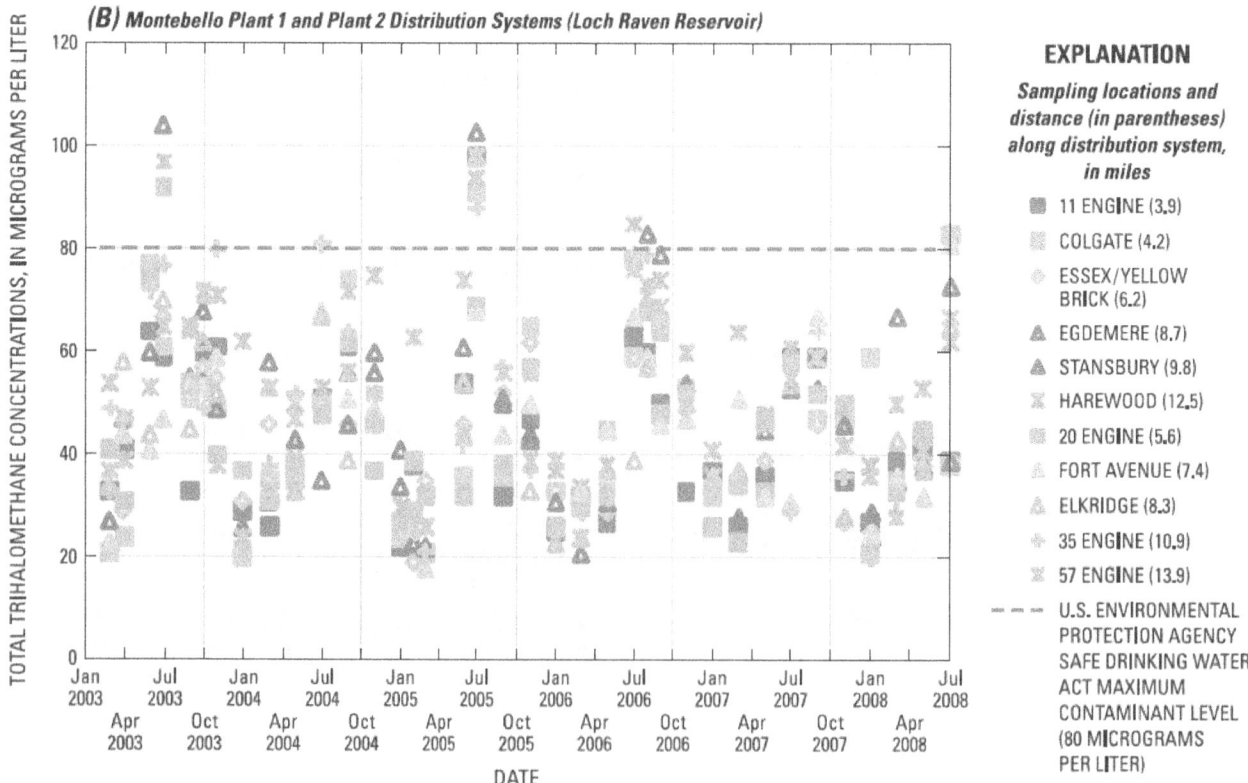

Figure 20.   Total trihalomethane concentrations at selected sampling locations in the water-supply distribution system from the
(A) Ashburton (Liberty Reservoir) and *(B)* Montebello (Loch Raven Reservoir) treatment facilities, January 2003–July 2008.
(Data provided by the Baltimore City Department of Public Works, Environmental Division, Baltimore, Maryland).

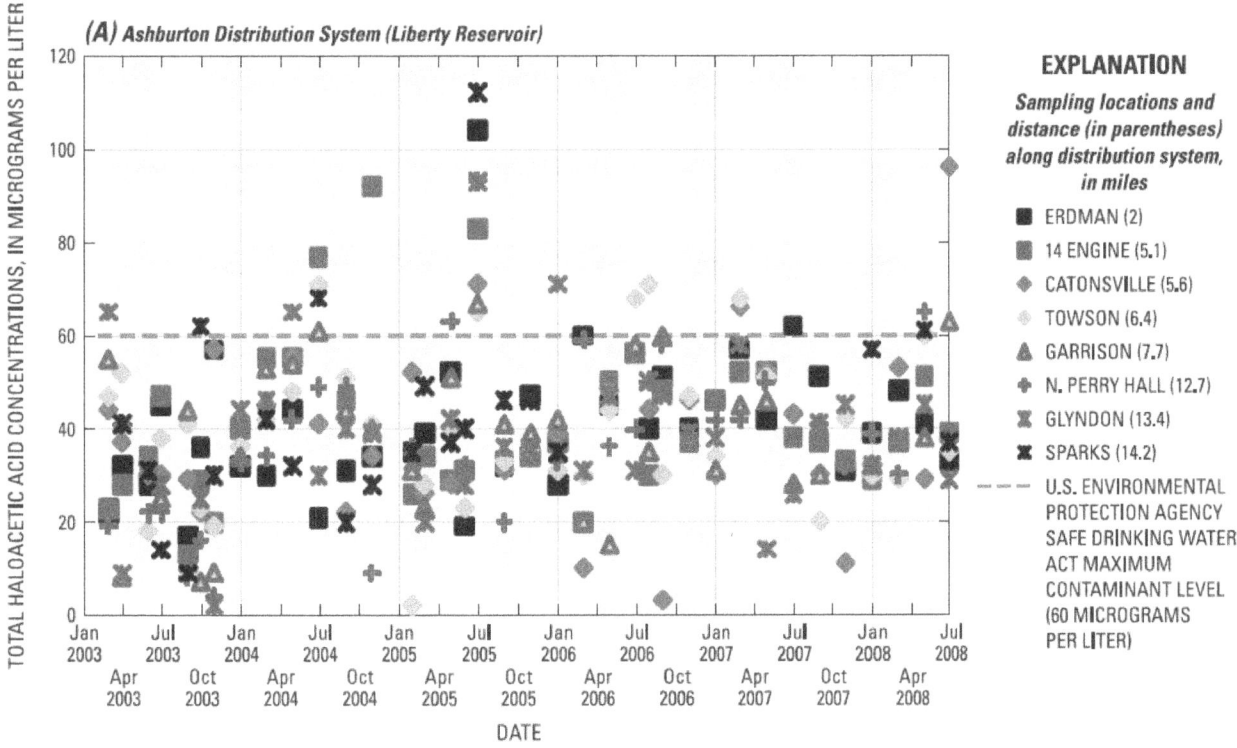

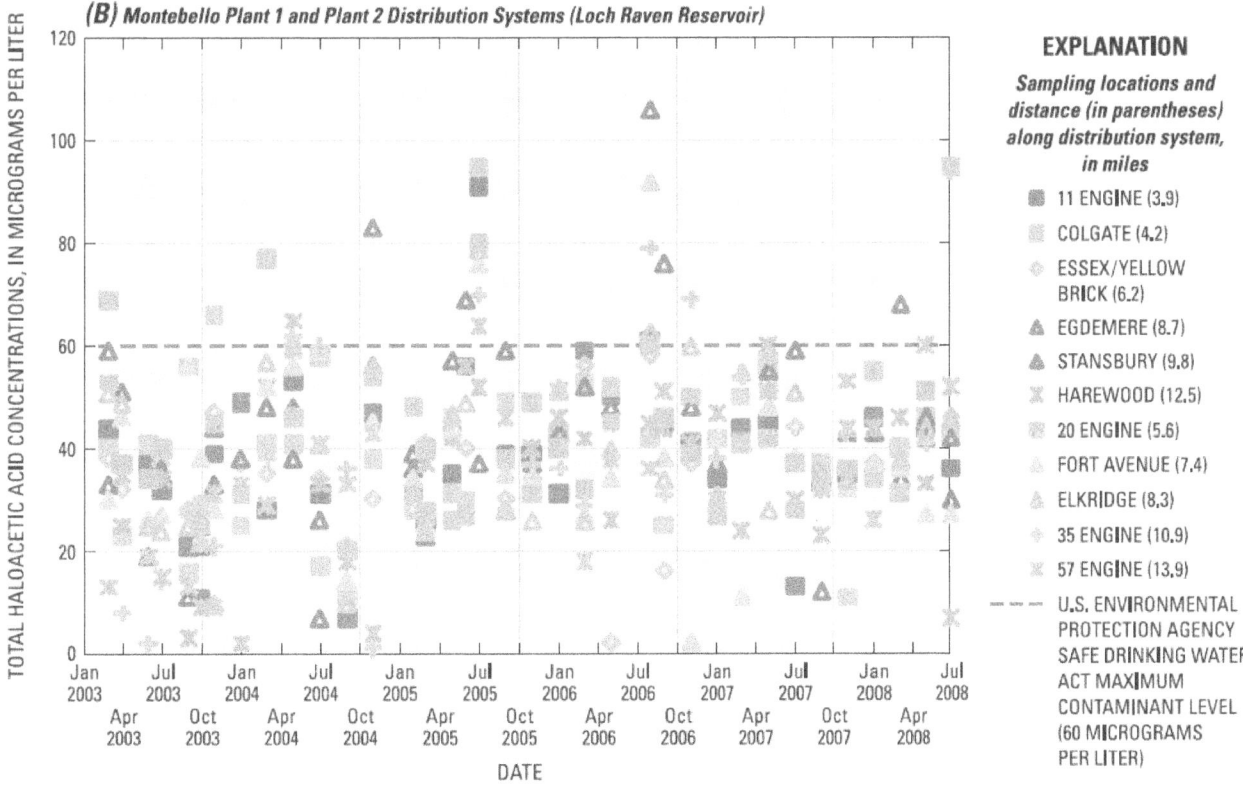

Figure 21. Total haloacetic acid concentrations at selected sampling locations in the water-supply distribution systems from the *(A)* Ashburton (Liberty Reservoir) and *(B)* Montebello (Loch Raven Reservoir) treatment facilities, January 2003–July 2008. (Data provided by the Baltimore City Department of Public Works, Environmental Division, Baltimore, Maryland).

Table 11.   Number of sampling dates that U.S. Environmental Protection Agency Maximum Contaminant Levels for disinfection by-products were exceeded at one or more stations in the water distribution systems at the Ashburton or Montebello treatment facilities, 2003–08.

[µg/L, micrograms per liter]

	Treatment facility	
	Ashburton	**Montebello**
Number of sampling dates that concentration of trihalomethanes equalled or exceeded 80 µg/L at one or more stations	7	7
Total number of sampling dates	37	37
Frequency of exceedances, in percent	18.9	18.9
Number of sampling dates that concentration of haloacetic acids equalled or exceeded 60 µg/L at one or more stations	16	14
Total number of sampling dates	37	37
Frequency of exceedances, in percent	43.2	37.8

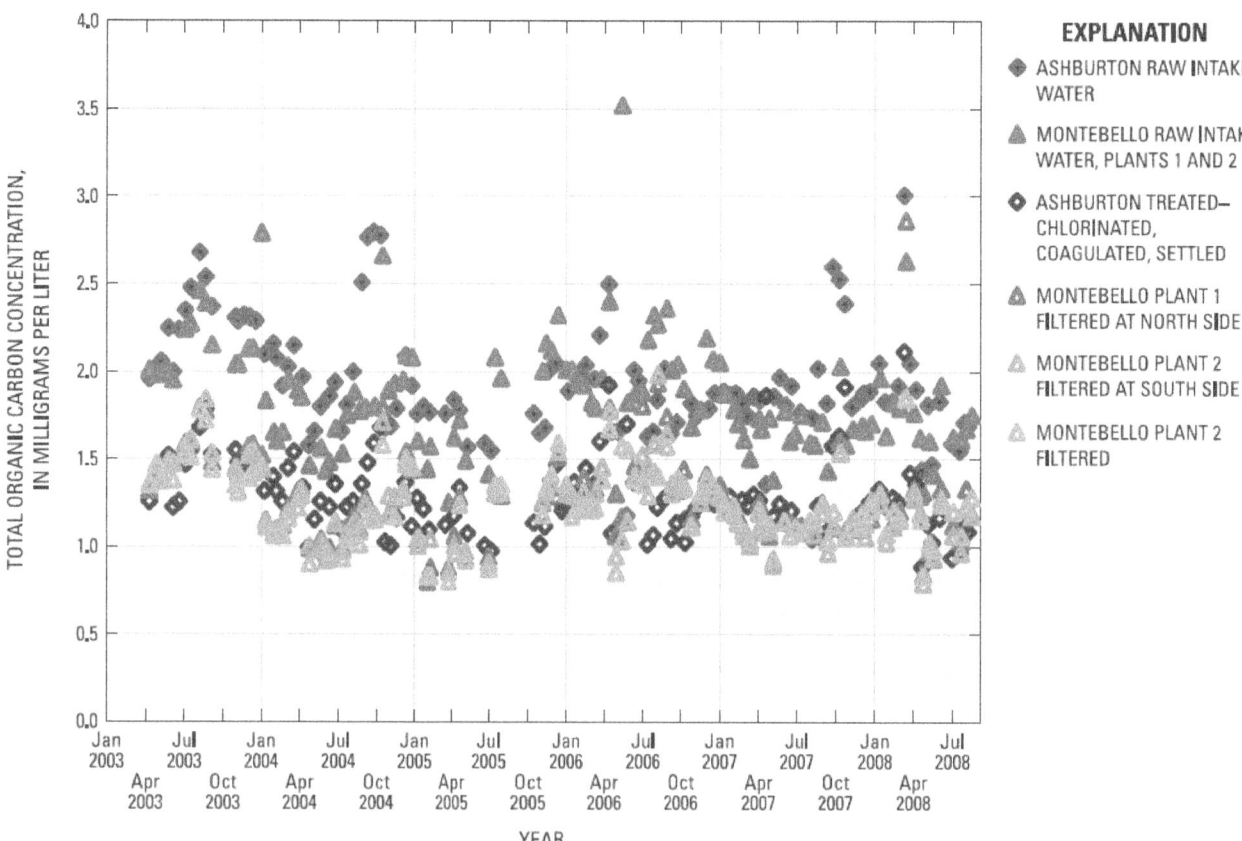

Figure 22.   Concentrations of total organic carbon in the Baltimore Reservoir treatment facilities, April 2003–July 2008 (Data provided by the Baltimore City Department of Public Works, Baltimore, Maryland).

Temporal variations did occur in concentrations of TOC during the period of record (fig. 22). At the Ashburton facility, the highest TOC concentrations occurred from mid-2003 to early 2004, which reflects reservoir recovery waters after the severe drought from 2000–02. Elevated TOC concentrations also occurred during the reservoir recovery period in late 2005–early 2006, and following a less severe drought in early 2005. These elevated concentrations of TOC possibly could reflect increases in carbon from the watershed tributaries in recovery flows, and (or) from algal blooms, or the decomposition of terrestrial vegetation established in the reservoirs during the droughts. Elevated TOC concentrations also occurred in the early fall in 2004 and in 2008. The elevated concentrations of TOC in 2004 again could reflect the continuation of increased levels of carbon in tributary flows, the decomposition of algal blooms, and (or) the decomposition of terrestrial vegetation established on reservoir bed sediments during the drought but then submerged during reservoir recovery. Concentrations of TOC in the Montebello facility, which generally reflect Loch Raven Reservoir water, but also occasionally reflect Susquehanna River water (see **Reservoir Watershed Management and Reservoir Operation**, *this report*), also appear to vary temporally throughout the period of record. Elevated TOC concentrations occurred during the drought-recovery periods noted above, as well as in the fall or winter—January 2004, October 2005, December 2006, and October 2007 (fig. 22). As was the case with Liberty Reservoir and the Ashburton facility, the periods of elevated TOC concentrations in the Montebello facility could, in part, reflect increases in carbon from Loch Raven Reservoir watershed tributary flows, reservoir algal blooms, and (or) the decomposition of terrestrial vegetation in the reservoir, during the recovery from drought conditions. The other periods of elevated TOCs, during the fall and winter months, could reflect increased carbon loads from the reservoir watershed tributaries, for example, leaf-litter contributions (October months) or lake turnover (January–December months), respectively. The variations in TOC at this facility in part also could reflect the use of Susquehanna River water—for example, during the severe drought and recovery from drought in 2000–04, and the moderate drought-recovery period from 2005–06 (see **Reservoir Watershed Management and Reservoir Operation**, *this report*).

Temporal variations in the concentrations of TOC were compared to either total THM or total HAA concentrations for the period of record. Although the monitoring of TOC in relation to THM and HAA concentrations was not specifically designed to relate the former to the latter, such comparisons are useful for illustrative purposes. In the case of the Ashburton facility, and except for 2003–04, the concentration of TOC bore no apparent relation to the total concentration of THMs during the period of record (fig. 23). Similar results were obtained for TOC and HAAs at this facility and for TOC in relation to THMs or HAAs at the Montebello facility. Variations in the amount and (or) types of carbon that form DBPs possibly are insufficient in magnitude to notably

influence the concentrations of TOC, or if there is such a correspondence, it is not apparent because of the limitations inherent in trying to compare these TOC and DBP data (see a) through e), p. 61).

The Baltimore Reservoir Technical Group (2004) recognized the need to determine the sources of carbon (rather than TOC) entering or generated within a reservoir that contribute to DBP formation. As such, future monitoring for in-stream and in-lake sources of carbon that form DBPs needs to take into account that the differences in the temporal patterns of elevated THMs or HAAs in finished waters in the water-supply distribution system could reflect different carbon sources.

Assuming that the variations in THM concentrations reflect variations in the concentrations and types of reactive carbon that form THMs, and not variations in treatment plant operations, freshly created TOC from phytoplankton can have a high propensity to form DBPs, and phytoplankton residues are not necessarily removed during coagulation and filtration (Graham and others 1998; Jack and others, 2002; Kraus and others, 2011). Elevated THM concentrations in finished water in the water-distribution system from the Ashburton facility and Liberty Reservoir generally do appear in mid-summer to early fall in most years of record, which corresponds with the period when algal blooms and die-offs most often occur in the reservoir. A similar pattern generally holds true for seasonal patterns in THM occurrence associated with the Montebello distribution system. In the case of this distribution system, however, the source of the finished waters generally is from Loch Raven Reservoir; but when reservoir water levels are low, the source of the finished waters could be the Susquehanna River.

In relation to major hydrologic events and sources of DBP precursors, and in the case of THMs, the drought-recovery periods (in 2003–04 and again in 2005–06) correspond to periods when relatively high THM concentrations occurred in finished waters at selected stations in both the Ashburton and Montebello distribution systems. These types of drought-recovery periods are known or suspected to be marked by intense algal blooms (Baltimore Reservoir Technical Group, 2004; Valcik, 1975). These blooms could result from tributary waters enriched in nutrients (Kaushal and others, 2008), and possibly also enriched in organic residues that contribute to DBP formation. In addition, terrestrial vegetation established on extensive areas of reservoir bed sediments exposed by repeated drawdowns during drought conditions (fig. 19) constitutes another potential DBP carbon source when this vegetation is submerged during reservoir recovery (Fram and others, 2001).

In the case of HAAs, elevated total HAA concentrations do not generally co-occur with elevated total THM concentrations in finished waters associated with either treatment plant, nor do they generally co-occur with elevated TOC concentrations in finished or treated waters. They also tend to occur at the same stations, but at different times in different years. These findings indicate that the sources and types of carbon that lead to elevated HAA concentrations could differ from the

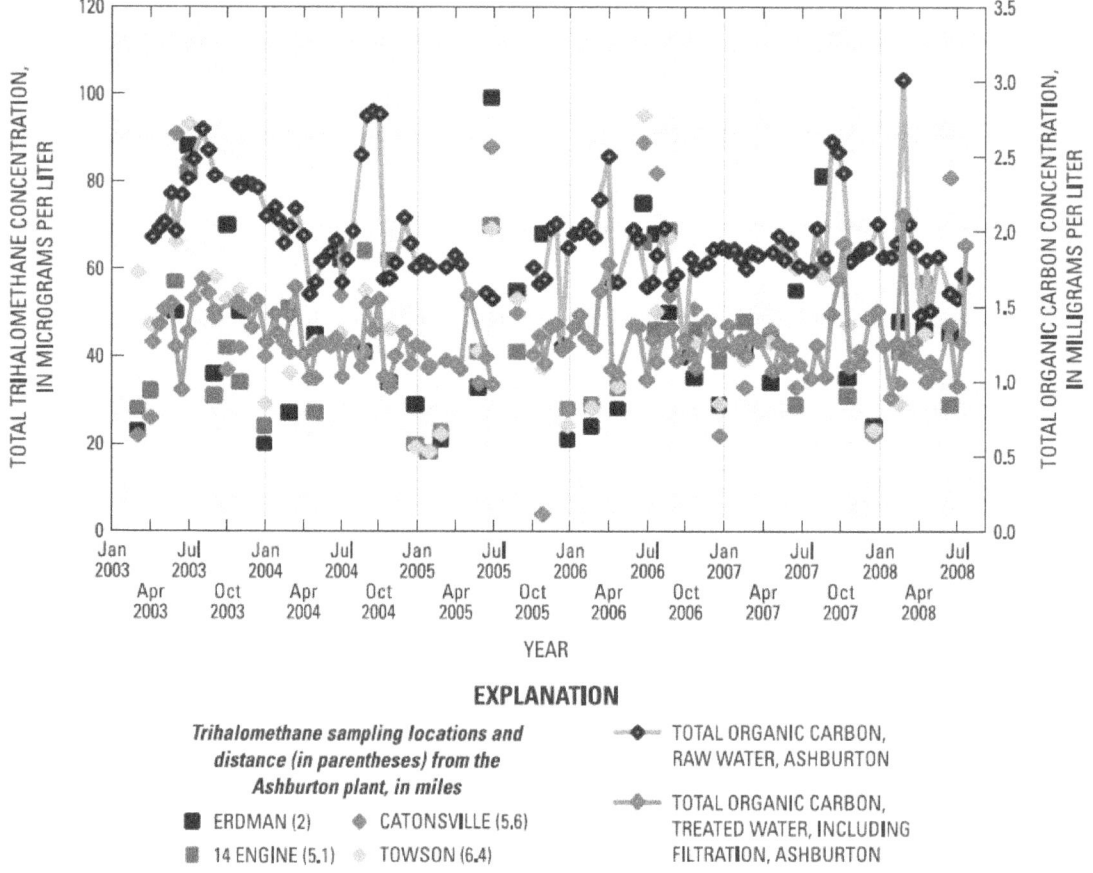

Figure 23.   Total concentrations of trihalomethanes for selected sampling locations in the Ashburton distribution system, and total organic carbon concentrations in raw intake and treated waters in the Ashburton treatment plant, April 2003–July 2008 (Data provided by the Baltimore City Department of Public Works, Baltimore, Maryland).

sources and types of carbon that form THMs, and, as in the case of THMs, are not of sufficient magnitude in concentration to lead to elevated concentrations of TOC. The sporadic nature of observed elevated concentrations of HAAs suggests these different sources and types of carbon are associated with relatively short and discrete biological and (or) hydrodynamic events whose nature is largely unknown.

In relation to future monitoring to determine in-stream and in-lake sources of reactive carbon that form DBPs, consideration needs to be given to the effect of the treatment process on DBP formation. The DBPs that form during initial chlorination of raw intake waters may or may not reach the distribution system. For example, research has shown that aromatic organic compounds more readily form DBPs than non-aromatic compounds, but that aromatic compounds often are effectively removed by coagulation (Fram and others, 2001, Singer, 1999; Wu and others, 2000.

In addition, future monitoring could be guided by a number of recent and ongoing studies. The USEPA rule change on DBPs has generated considerable interest, discussion, and study to identify the sources and types of carbon that lead to DBPs in finished waters (Bergamaschi and others, 1999; Fram and others, 2001; Mash and others, 2004; Xie, 2004, Bergamaschi and others, 2005; Kraus and others, 2008; Kraus and others, 2010; Kraus and others, 2011). On the basis of these studies, the following findings could be relevant to the Baltimore reservoir system and City water-treatment process, and help guide monitoring to identify the sources of carbon that form DBPs within the Ashburton and Montebello water-distribution systems:

a) DBP carbon sources can be autochthonous—for example, from reservoir phytoplankton or aquatic plants, and (or) allochthonous—for example, from humic, silt, or colloidal detritus (plant or animal) matter transported by tributaries.

b) Autochthonous sources could include vegetation initially established on bed sediments during a drought, and subsequently submerged during reservoir recovery.

c) The source of carbon that leads to DBP formation could be allochthonous, and not be found to form DBPs if exposed to chlorination; but this carbon could undergo transformations as it is transported through the reservoir in a manner that enhances its ability to form DBPs.

d) Differences in the physical and chemical characteristics of autochthonous and allochthonous source materials can aid in their identification as DBP precursors.

e) The types of DBPs that enter, or form afterwards in, the distribution system depend upon the water-treatment process; variations in that treatment process—including levels of initial chlorination, coagulation and filtration; the length of time finished water is stored, and its manner of storage; the re-chlorination of finished water before it is placed in the distribution system; and the amount of time finished water resides in the distribution system.

f) Accurate assessment requires that potential DBP source waters be sampled and tested using treatment and post-treatment processes with residence times similar to that used when treating and distributing water for consumptive use.

g) Reducing DBP formation from a confirmed source, often requires determining at what point(s) in the treatment process that source forms DBPs, and how quickly the DBPs form and decompose.

h) In areas with heavy use of road salt (sodium chloride) as a deicing agent, bromide (a residual contaminant in road salt) could contribute to elevated DBP concentrations as brominated DBP compounds have an appreciably greater mass than chlorinated DBP compounds.

Given all of the above factors, continuous-monitoring parameters and procedures are being identified and developed to help identify DBP precursors in source waters, which effectively monitor for more than TOC. Effective monitoring generally (a) begins with the determination of concentrations of aromatic and aliphatic carbon in raw reservoir intake waters, (b) follows with treatment of samples to determine the relative contributions of aromatic and aliphatic carbon to DBP formation, (c) extends to backtracking to identify reactive carbon source(s) in the reservoir and its tributaries, and (d) extends to forward tracking of these reactive carbon compounds to determine if these compounds occur in proximity to where the DBPs form in the distribution system.

Sodium and Chloride

Elevated concentrations of sodium and chloride in potable water supplies are emerging water-quality concerns for the RWMA partners. Since the early 1970s, when sodium concentrations began to be routinely measured, concentrations of sodium have almost tripled in the supply-intake water from Liberty Reservoir, and almost quadrupled in the supply-intake water from Loch Raven Reservoir (Baltimore Reservoir Technical Group, 2004). In 1999 and 2003, sodium concentrations in treated waters from Loch Raven Reservoir often exceeded 20 mg/L—the USEPA (2003) non-regulatory health advisory for consumers on restricted low-salt-intake diets (500 mg/d or milligrams per day). During the same period, sodium concentrations in treated waters from Liberty Reservoir ranged from 10–16 mg/L, exceeding almost all previous measurements.

Chloride concentrations also have increased in reservoir intake waters since the early 1970s from about 15 mg/L to as high as 25 to 30 mg/L by the year 2000. Although still well below the USEPA SWDA health advisory (2009), which is 250 mg/L, the increase in chloride concentrations has raised RWMA partner concerns, as indicated in their most recent RWMA 2005 (see **Reservoir Watershed Management and Reservoir Operation**, *this report*).

In relation to monitoring, sodium has not been extensively and continuously monitored by the City. Although routine monitoring for sodium began in 1973 in finished waters at the intakes of both water-supply reservoirs, the data largely remained unpublished until the 1990s. Up to this period, sodium concentrations were well below the USEPA SDWA secondary standard for taste set for the general population (30–60 mg/d). Thus, it was only recently that MDE (2004a) provided some historical data on finished waters at the Montebello treatment facility (fig. 24). The data illustrate the rise in sodium concentrations described above by the RTG for finished waters from 1973–2004. As will be discussed in greater detail below, the source of the recent increase in sodium concentrations has been attributed to the use of salt (sodium chloride) as a deicing agent in high-density parking and transportation corridors (Reservoir Watershed Protection Committee, 2000; Baltimore Reservoir Technical Group, 2004). The basis for this conclusion is that recent sodium concentrations are roughly proportional to the increases in chloride concentrations and conductance levels—both of which have been monitored extensively and for long periods of time in the reservoir watersheds and (or) reservoirs.

From a historical perspective, chloride concentrations in the Montebello treatment facility intake waters were routinely monitored beginning soon after the reservoir was created. Data provided by Amatayakul, Defries, and others (1978) show that chloride generally increased from the 1930s to 1970s (fig. 25). The initial upward trend in chloride concentrations to the mid-1940s likely reflected runoff from agricultural development (livestock and animal waste) and human septic and sewered wastes. Following World War II, increases in urban residential,

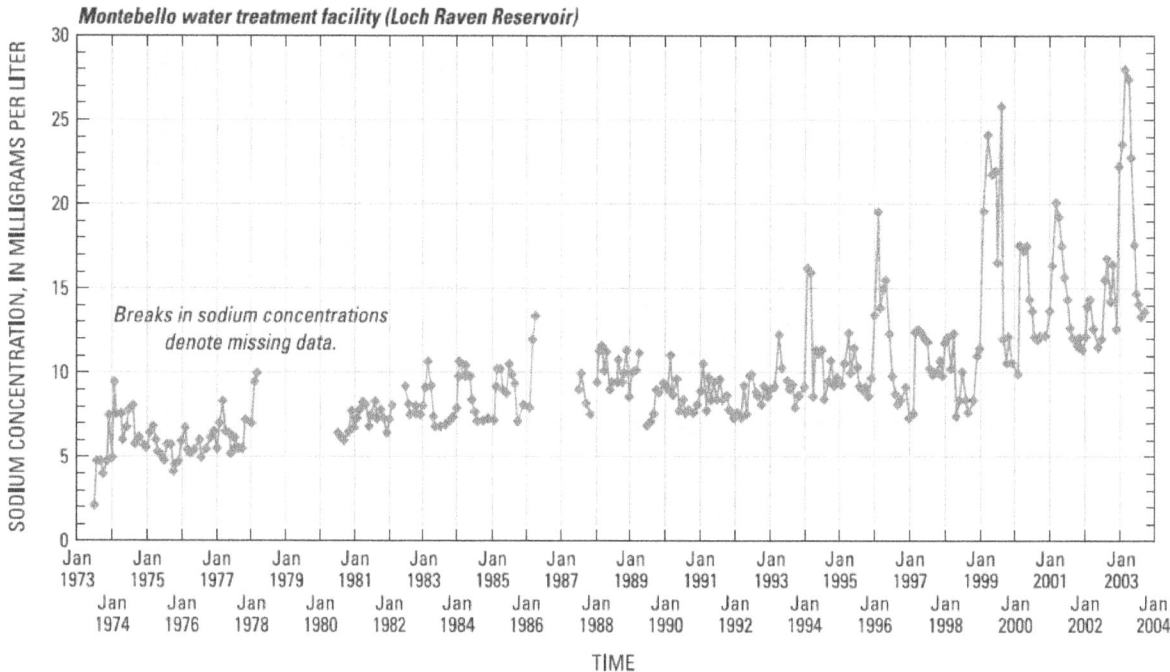

Figure 24. Concentration of sodium in intake waters at the Montebello water treatment facility, Loch Raven Reservoir, January 1973–January 2004 (modified from Maryland Department of the Environment, 2004a).

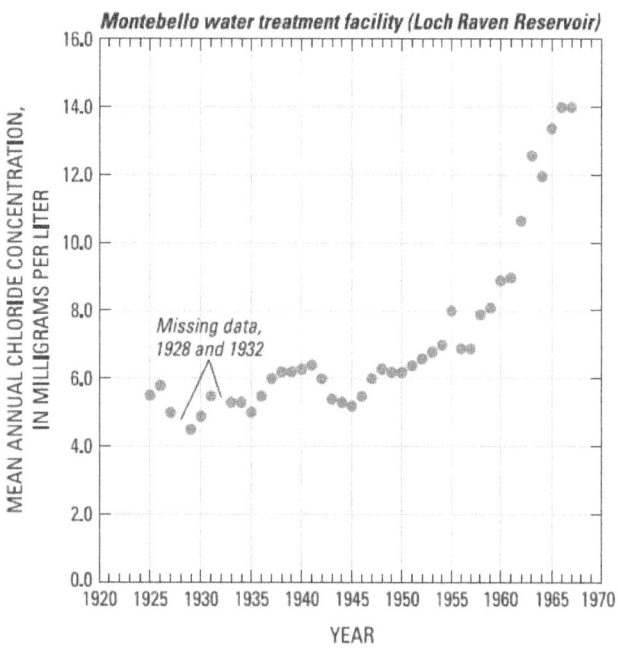

Figure 25. Mean annual chloride concentration in intake waters at the Montebello water treatment facility, Loch Raven Reservoir, 1925–67 (modified from Amatayakul and others, 1978).

commercial, and industrial growth likely contributed to the sharp rise in chloride concentrations that continued up to the mid-1960s. By then, and just before the first major sewage WWTPs came on line in the watershed (at Hampstead in 1969 and at Manchester in 1970), chloride concentrations at the reservoir intake had risen from approximately 5 mg/L in the 1930s to 14 mg/L.

From the mid-1970s through the 1980s, chloride monitoring was intermittent, and not performed as part of routine reservoir in-lake monitoring. Up until this period, it was primarily used as a surrogate for nutrient (primarily nitrate) concentrations; a use which was halted because of concerns that natural sources of chloride (for example, sea spray) rendered its use as a surrogate for nutrients suspect. In addition, analytical methods to directly measure nutrients became available in the late 1960s.

Since 1982–83, chloride and conductance data were again collected by the City at selected reservoir and watershed tributary stations. A comparison of selected time-series data for in-lake conductivity and chloride concentrations (fig. 26), as well as in-lake conductivity in the reservoirs and chloride concentrations in the upstream tributaries (fig. 27), indicate apparent positive trends in chloride and conductivity, and that the chloride concentrations in the reservoirs have continued to increase in reservoir waters (Baltimore City Department of Public Works, 1996, 2001; Winfield and Sakai, 2003).

Since 1992, dry-weather-flow data indicate chloride concentrations also apparently have increased in most, but not all, of the reservoir subbasins. For the Loch Raven and Prettyboy

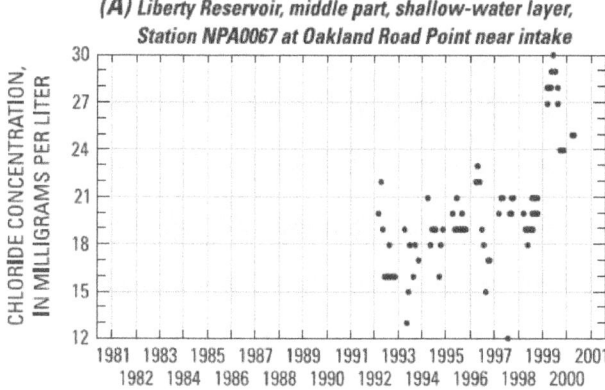

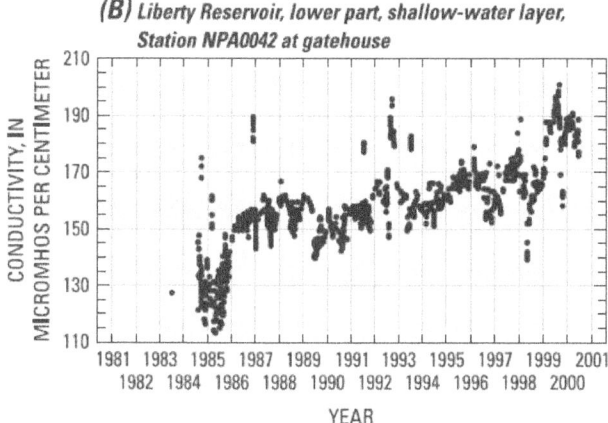

Figure 26. *(A)* Chloride concentrations, 1992–2000, and *(B)* conductivity, 1984–2000, of in-lake shallow waters (less than 30 feet deep) in Liberty Reservoir (modified from Baltimore City Department of Public Works, 2001).

Reservoir watersheds, and except for Beaver Dam Run, all the monitored subbasins show apparent increases in chloride concentrations; in the Liberty Reservoir watershed, and except for the stations at Beaver Run and the North Branch of the Patapsco River, all dry-weather-flow tributary stations also show apparent increasing trends (Baltimore City Department of Public Works, 1996; Winfield and Sakai, 2003).

Monitoring data have indicated that conductivity and chloride concentrations in the tributaries and shallow reservoir waters have varied seasonally, and have been subject to periodic spikes (figs. 24, 26, and 27). The highest conductivity and chloride concentrations typically occurred in the late winter or early spring, whereas the lowest conductivity and chloride concentrations typically occurred in the summer months (Baltimore City Department of Public Works, 1996).

Winter maxima in chloride concentrations appear related to subbasin land use. Using data from the 1999 water year, which had the highest winter chloride concentrations on record at supply-reservoir intakes, annual median chloride concentrations in dry-weather flows for subbasin tributaries

were found to be most significantly ($p < 0.0054$) correlated to the areal extent of commercial and industrial land use within a subbasin (fig. 28). The median chloride concentration in dry-weather flows for the winter (January through March) periods of 1996–99 also were found to significantly correlate with the road length per unit area ($p < 0.0313$) in a subbasin (fig. 28). In addition, the difference in the 1999 and 1990 median annual chloride concentrations in dry-weather tributary flows were significantly correlated ($p < 0.0116$) with one landscape characteristic—declining agricultural land use (fig. 28). Further evidence that urban settings possibly contributed to a disproportionately high amount of chloride in the reservoirs was provided by Winfield and others (2006). They found an increasing pattern in chloride concentrations from reservoir intakes to upstream urban areas (fig. 29).

Monitoring for chloride in the watershed tributaries and at the reservoir intakes, and to a lesser extent, for sodium in finished drinking waters, has enabled the RWMA partners to determine that the recent increases in sodium and chloride were cause for concern. As of 2007, monitoring has shown that sodium in reservoir intake waters has approached concentrations that occasionally exceed Federal guidelines for individuals on low-sodium intake diets. Chloride concentrations also have increased at rates similar to those observed for sodium. The analysis of recent apparent trends in the monitoring data with land use also has enabled the identification of road-salt use as a deicing agent, particularly in residential, institutional, commercial, and industrial areas with a high density of impervious parking areas and roads, as likely being responsible for the enhanced elevated concentrations of sodium and chloride in finished water supplies during the winter months.

As of 2007, monitoring cannot provide a timely advance warning to water purveyors (at the Ashburton and Montebello treatment facilities) of potentially high sodium concentrations. Sodium is not routinely or frequently measured in either the reservoirs or in the tributary watersheds. Nor does chloride sampling occur with sufficient frequency during the winter months to use as a surrogate for sodium. In general, monitoring for either sodium or chloride concentrations in winter months in both the reservoir watersheds and reservoirs also is insufficient to readily detect reduced concentrations should actions be taken to reduce sodium-chloride road-salt use in the subbasins. It could take years for such reductions to become truly apparent given the type and frequency of sampling being conducted. It also is unclear whether or not sodium and chloride concentrations will decline or simply increase at a reduced rate following a major reduction in road-salt use because there are other sources of sodium and chloride in the reservoir watershed. For example, watershed populations are expected to continue to increase, which either likely will increase sodium and chloride loads to WWTPs, or from additional septic systems, which in either case, do not remove sodium or chloride, and eventually likely lead to increased concentrations being observed in reservoir-watershed tributary streams.

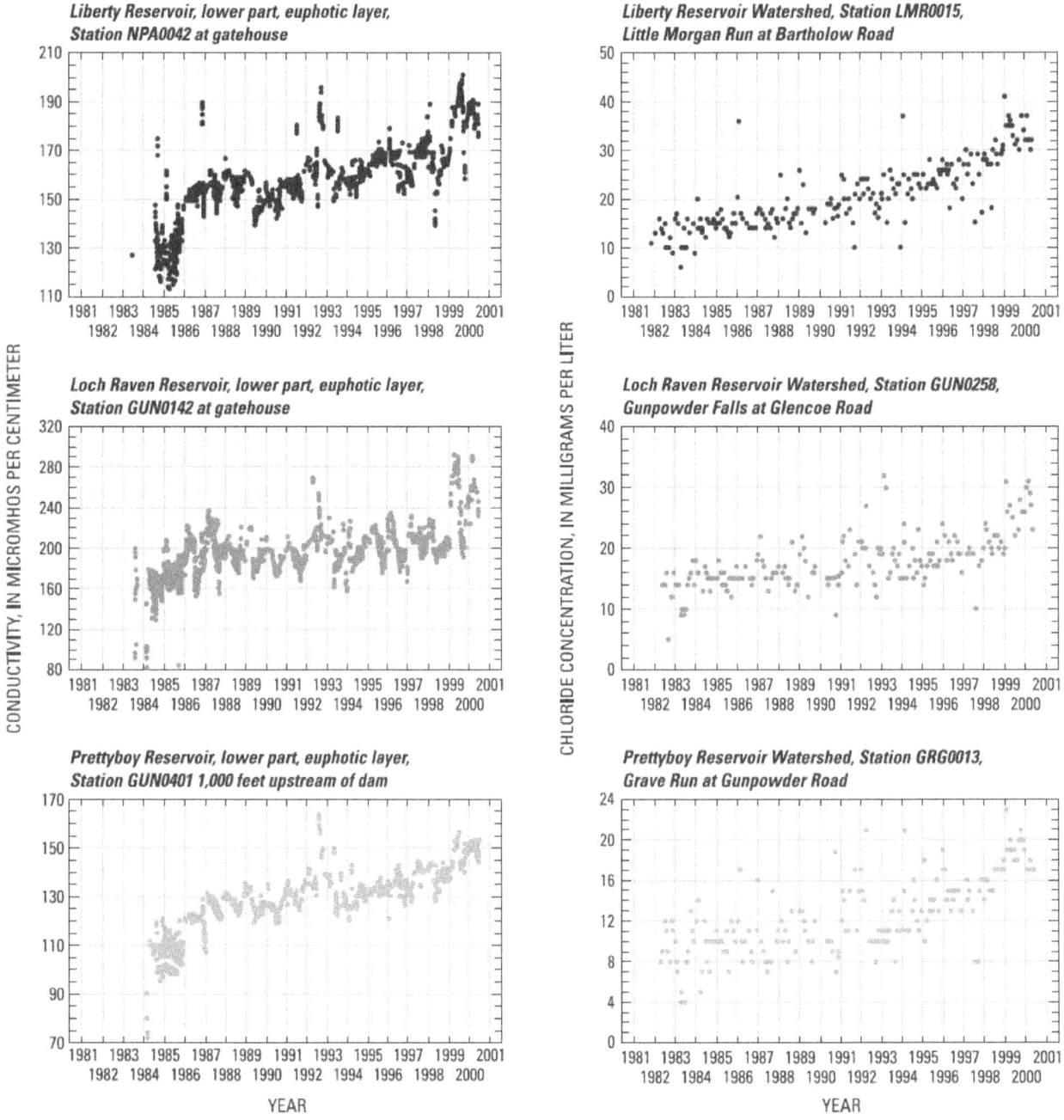

Figure 27.   Examples of conductivity in shallow waters (less than 30 feet deep) of the Baltimore Reservoirs, and chloride concentrations in upstream tributary dry-weather flows, 1982–2000 (modified from Baltimore City Department of Public Works, 2001).

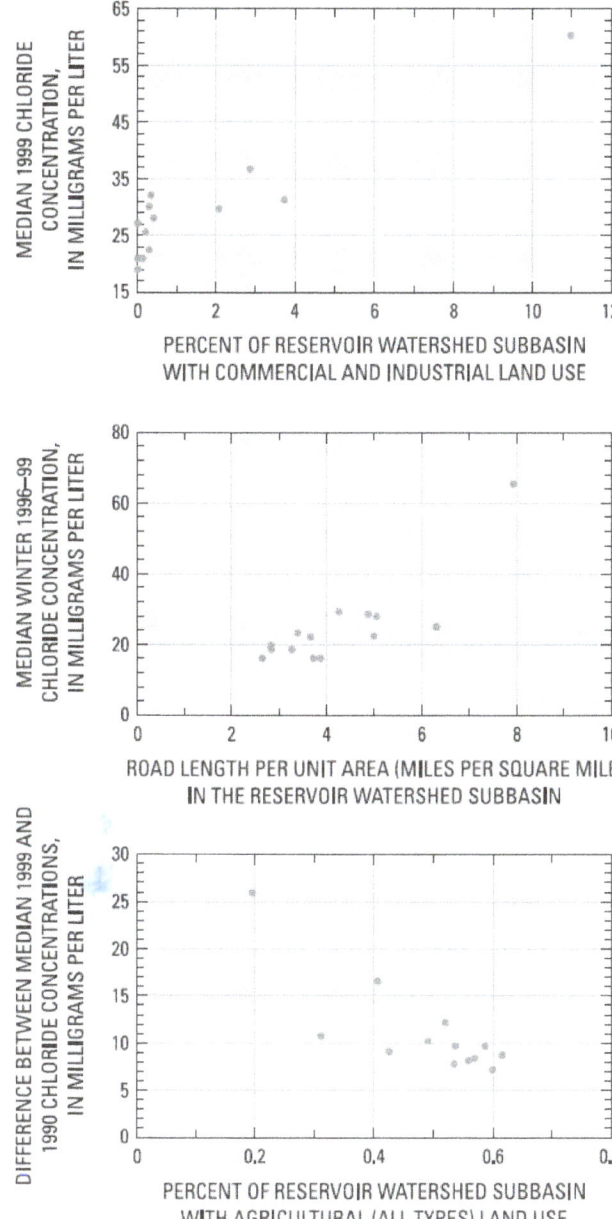

Figure 28. Relations between median concentrations of chloride in all dry-weather tributary flows for selected time periods and selected 1999 landscape characteristics in tributary subbasins of the Baltimore Reservoir watersheds, various years of record (modified from Baltimore City Department of Public Works, 2001).

Effects of Climate

Climate has a marked effect on the quality as well as quantity of water in the tributaries and reservoirs of the Baltimore Reservoir System. The climate of the Mid-Atlantic region over short time periods is inherently quite variable. This variability is apparent in the within-year and year-to-year variations in streamflows, tributary concentrations and loads, and reservoir water quality for most pollutants of concern (for examples, see **Eutrophication—Nutrients, Sedimentation—Sediment Transport,** and **Sodium and Chloride**, *this report*). In terms of short-term climatic variations, those that are possibly most critical are potential changes in water quality associated with major storms and droughts and recoveries from droughts (see **Overview of Water-Quality Concerns,** *this report*). Within the last decade, however, there have been at least two moderate to severe droughts that have affected reservoir water quality and operations, and that were largely terminated by major storms—a pattern that could reflect recent long-term forecasts of the effects of climate change on the Mid-Atlantic region. As of 1995, however, the monitoring program does not appear to be well suited to ensuring that the effects of climate variability and climate change on the quality of water in the watershed tributaries and reservoirs can be clearly assessed and differentiated from the effects of watershed and reservoir management activities designed to address RWMA water-quality concerns.

Climate Variability

Whereas year-to-year and within-year variations in tributary and reservoir water quality inherently have been characterized by the monitoring program to a degree, there are two aspects of climate variability—storms and droughts—for which the effects on watershed tributary and reservoir water quality have not been well characterized.

Monitoring of stormflows declined in the mid-1990s, and as of 2007, and except perhaps for drought years, the subsequent monitoring of stormflows within a year and from year to year likely has been insufficient to characterize stormflow water quality and pollutant loads of interest (see **Eutrophication—Nutrients: Phosphorus Transport and Reservoir Recycling** and **Sedimentation—Sediment Transport,** *this report*). However, even the storm-related data collected before the decline in stormflow monitoring possibly were insufficient as suggested by the most recent and advanced attempt to characterize the effects of tributary flows and water quality on reservoir water quality and biota (see **Modeling to Address Water-Quality Concerns,** *this report*). According to the ICPRB, there appears to be insufficient data to determine if the effect of storms on reservoir water and biotic quality is simply minimal and transient (Interstate Commission on the Potomac River Basin, 2006). Historical monitoring of tributary stormflows, and in the reservoirs, has been too infrequent. Reservoir monitoring at selected sites also is limited by boat and weather permitting (see **Current Perspective,** *this report*).

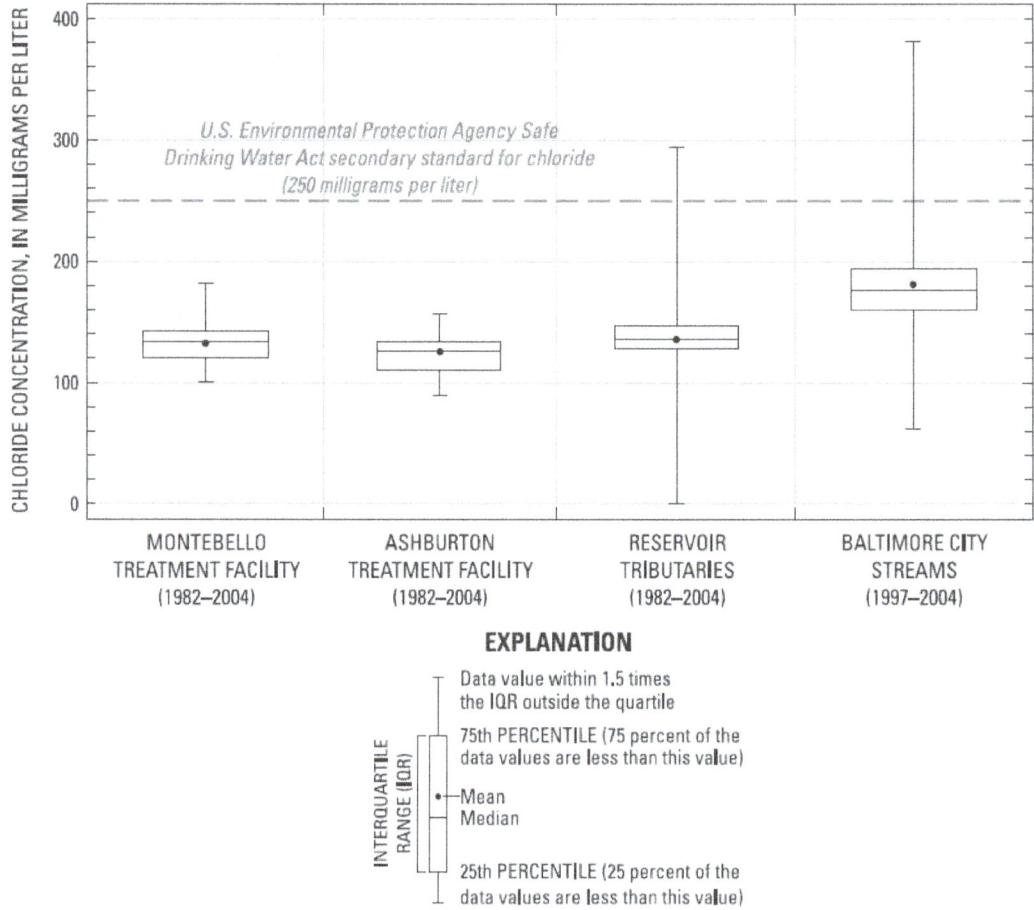

Figure 29. Chloride concentrations for finished supply waters at Montebello and Ashburton treatment facilities (1982–2004), reservoir tributaries (1982–2004), and selected Baltimore City streams (1997–2004) (modified from Winfield and others, 2006).

The effects of droughts and recoveries from droughts represent another form of short-term climatic variability for which the effect on watershed tributary quality, and reservoir water and biotic quality, also is not well known. Anecdotal evidence provided during the late 1960s drought and 1970s recovery (Valcik, 1975), and also more recently by Winfield and Sakai (2003), the Baltimore Reservoir Technical Group (2004), and MDE (2004a), for the 2000—03 drought-with-drawal-recovery event, indicates that algal blooms could be enhanced during a major drought and (or) during the recovery from drought conditions. Reservoir withdrawals during pro-longed droughts could be a contributing factor because of the City firming program which aims to always withdraw the best quality of water (see **Reservoir Watershed Management and Reservoir Operation**, *this report*).

Beginning in the summer of 1999 until the spring of 2003, the reservoir watersheds experienced moderate and then severe drought conditions marked by episodic periods of relatively low streamflow (fig. 30). Although precipitation,

such as the nor'easter in the winter of 1999, which resulted in a major snowfall in the region, was initially sufficient to reduce the impact of the drought, drought conditions intensi-fied and persisted until 2003. Final cessation of the drought and full recovery in the reservoirs began with Hurricane Isabel in September 2003, and was enhanced by two nor'easters that produced major snowfalls in the region in December 2003 and February 2004.

During the drought-recovery period, and due to the City firming program, it is evident that annual releases from Prettyboy Reservoir increased in magnitude each year until drought recovery (fig. 31), and in response to major withdraw-als and drawdowns each year from Liberty and then Loch Raven Reservoirs, in the absence of full reservoir recovery. Ultimately, the City obtained water from the Susquehanna River (see **Reservoir Watershed Management and Reservoir Operation**, *this report*), but not before reservoir withdrawals led to parts of each reservoir being drained, and covered by terrestrial vegetation (for example, see fig. 19).

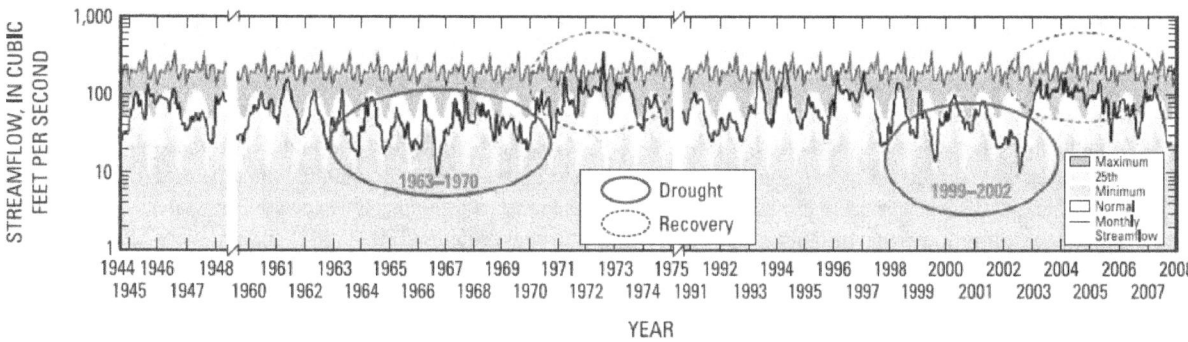

Figure 30. Long-term streamflow characteristics with major drought-recovery periods for U.S. Geological Survey station LIT0002 (Little Falls at Blue Mount, Maryland) in the Loch Raven Reservoir watershed, 1941–2007 (modified streamflow records from U.S. Geological Survey Maryland-Delaware-D.C. Water Science Center).

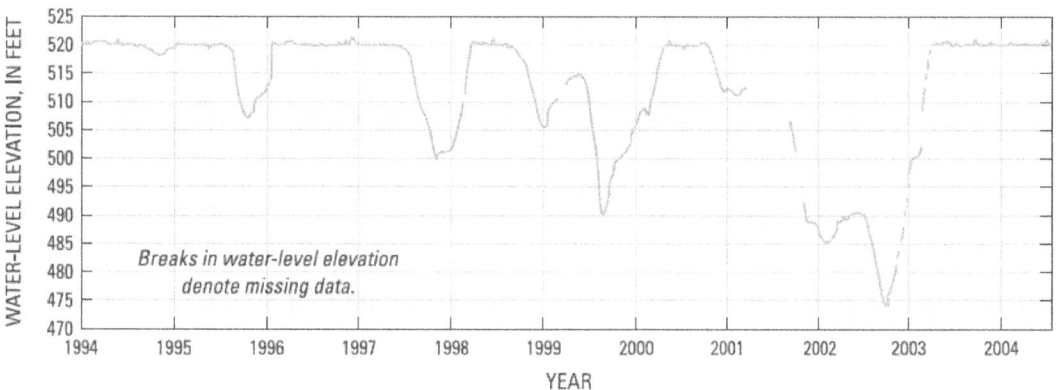

Figure 31. Annual variation in surface-water elevation in Prettyboy Reservoir, indicative of releases to replenish Loch Raven Reservoir under normal and drought conditions, 1994–2004 (modified from Interstate Commission on the Potomac River Basin, 2006).

Upon submergence during reservoir recovery, the decomposition of this vegetation potentially could have led to elevated concentrations of mercury methylation and biological uptake in the reservoirs and elevated concentrations of DBPs in drinking water (see **Sedimentation—Sediment Diagenesis and Mobilization of Metals and Phosphorus** and **Disinfection By-Products**, *this report*). In addition, the absence of large volumes of water in the reservoirs likely contributed to the elevated concentrations of sodium and chloride that occurred in reservoir supply intake waters in 1999 and 2003–04 (Maryland Department of the Environment, 2004a). The source of this sodium and chloride is likely from road-salt use associated with the major snowfalls produced by the nor'easters that occurred in December 1999, December 2003, and February 2004 (see **Sodium and Chloride**, *this report*).

There also may have been no prolonged recovery as another moderate drought and recovery occurred in 2005–06. During this drought-recovery period, however, reservoir withdrawals initially were reduced by taking water from

the Susquehanna River before major declines had occurred in reservoir water levels. However, the poor quality of this river water led to a decline in the quality of treated water, and its use was discontinued (see **Reservoir Watershed Management and Reservoir Operations**, *this report*). Fortunately, the drought was short-lived, with recovery in 2006 being aided by Tropical Storm Ernesto (September 2006) and two back-to-back nor'easters in November 2006.

During this review, no studies were encountered that specifically analyzed changes in reservoir watershed tributary, or reservoir biotic and water-quality conditions associated with major storms or drought-drawdown-recovery events in the Baltimore reservoir system during the past decade. Nor does the City firming program appear to have been reviewed to determine whether or not the extended withdrawals of the best available water affect the determination of subsequent reservoir water quality during recovery from drought conditions.

It is possible that reservoir water-supply systems such as the Baltimore reservoir system are particularly vulnerable to extended periods of dry weather followed by reservoir recovery aided by intense storms. During droughts there are increased demands for water supplies, particularly during extended dry periods that include the spring, summer, and early fall. Under the City firming program, initial drawdown of either supply reservoir is somewhat mitigated, but under extended dry periods with continuing demands, reservoir withdrawals eventually remove major quantities of the best quality water from one and progressively both water-supply reservoirs, with the remaining water possibly trending towards poorer quality. Continued withdrawals also ultimately produce major declines in reservoir volumes, and possibly in the quality of water that remains. Thus, recovery, even if temporary, occurs with increased tributary inflows, likely enriched in nutrients and sediment, into reservoirs with reduced dilution volume and buffering capabilities, creating conditions favorable for algal proliferation.

Documentation of the individual or combined effects of major storms and droughts, withdrawals, and recoveries on water quality in the reservoirs is lacking. The historical and current monitoring program appears to lack sufficient spatial and temporal resolution to accurately characterize intra-seasonal changes in reservoir biotic and water-quality conditions that likely result from these types of events (for example, see **Modeling to Address Water-Quality Concerns**, *this report*).

Climate Change

The last decade could be an indicator of the future effect of climate change on the Baltimore Reservoir system given that the forecasted regional effects of climate change are an increase in the frequency of moderate-to-severe droughts punctuated by an intense increase in storms. As of 2007, detailed analysis of long-term monitoring and modeling results provided by the Intergovernmental Panel on Climate Change (IPCC) predicts future increases in both precipitation and hydrologic extremes for the Mid-Atlantic region (Intergovernmental Panel on Climate Change, 2007). The IPCC has concluded that the frequency of drought conditions is likely to increase (66–90 percent probability) in the Mid-Atlantic region and that these conditions are very likely (90–99 percent probability) to be interspersed with periods of heavy precipitation—for example, due to tropical storms and nor'easters as described above. Assuming climate-change predictions are accurate, and depending upon the climate scenario adopted, recent studies indicate that Maryland could receive about 10 percent more precipitation in winter and spring (Maryland Department of the Environment, 2008). Because of warmer temperatures overall, however, it was noted that rainfall could be more intermittent, extreme events more likely, and drought durations of several weeks to months in duration could be common during what amounts to an extended and dry "summer" season.

The effects of these climate-related changes, combined with the accelerated conversion of land to human-dominated uses (Vitousek and others, 1997) could result in notable increases in contaminant loads to reservoir-tributary rivers and streams in the form of intense storm events. Kaushal and others (2010) documented the extent of this climate-land use interaction in the Chesapeake Bay watershed, where the extreme drought (1999–2002) and recovery (2003) ultimately led to near record freshwater inflows to the Bay. The result was an increased flux of nitrogen to the Bay that coincided with phytoplankton blooms, severe hypoxia, and elevated mortality of fish and other aquatic species (Acker and others, 2005; Miller and others, 2006). Kaushal and others (2008; 2010) concluded that nitrogen stored in tributary watersheds during the extended dry period was flushed into the waterways by the intense rainfall and runoff. They also found that agricultural, and in particular urban, areas had accelerated rates of nutrient transport and reduced retention capacities, to the extent that, even after a return to more typical streamflow conditions in 2004, the flux of nitrogen remained high.

Other studies have further highlighted the combined effects of development, and in particular, urbanization, and increased climate variability on water resources in the eastern United States. Langland and others (2007) provided evidence that sediment loads in the Potomac River near Washington, D.C. could be increasing in response to increased variability in streamflow since the 1970s. Wall and others (2008) showed that recent high-flow events in the Hudson River estuary produce elevated inputs of sediment over short time intervals. In addition to increased nutrients and sedimentation, drinking-water supply reservoirs and rivers in the region could be especially vulnerable if climate change also results in increased snowfall. The latter would require increased applications of road deicing salts in urban and suburban areas, which could possibly increase pulses of sodium and chloride (Kaushal and others, 2008).

Kaushal and others (2010) have indicated that increases in the frequency and magnitude of contaminant pulses will require major changes in existing water-quality monitoring programs, including the need for high-resolution, continuous and (or) high frequency monitoring and event sampling. Continuous monitoring of specific conductance, for example, is a relatively inexpensive surrogate for sodium and chloride measurements (Granato and Smith, 1999); and increased episodic or event sampling, although costly, is essential to quantifying the effects of intense precipitation on contaminant loading (Lindenmayer and Likens, 2009; Palmer and others, 2008).

The monitoring requirements described above differ markedly from the trend in monitoring in the Baltimore Reservoirs and their contributing watersheds since the mid-1990s and at least through 2007, which generally has emphasized monthly dry-weather tributary sampling and intermittent monthly to bimonthly reservoir monitoring. Major sediment and nutrient-laden stormflows generally are not adequately monitored. However, they can reach one of the Baltimore Reservoirs in a matter of hours (see **Appendix A—Watershed Characterizations**, *this report*). Their effect on reservoir hydrodynamic, water-quality, and biotic and conditions is not well understood. Their initial arrival and subsequent impact likely often go undetected because of the relatively low (monthly or bimonthly) frequency of monitoring in the reservoirs (see **Modeling to Address Water-Quality Concerns**, *this report*), which is further limited by the inability to access selected monitoring sites except by boat and weather permitting (see **Current Perspective**, *this report*).

Integrated Framework for an Enhanced Water-Quality Monitoring Program

The design and evaluation of an integrated framework for a comprehensive water-quality monitoring program begin with a clear definition of goals and objectives (Reinelt and others, 1988). The goals guide the development of the entire monitoring-program process, including its design, implementation, and evaluation. Ideally, the data obtained effectively meet the monitoring objectives. An effective water-quality monitoring program typically provides answers to the following types of questions (Intergovernmental Task Force on Monitoring Water Quality, 1995): What is the condition of the source water? Where, how, and why are water-quality conditions changing over time? What problems are related to source-water quality? Where are these problems occurring, and what is causing them? Are programs to prevent or remediate problems working effectively? Are water-quality goals and standards being met?

Specific goals of the RWMA partners and objectives of the RTG incorporate these questions as follows:

a) To characterize the condition of the water supply with regard to contaminant occurrence, sources, causative agents, and potential threats.

b) To detect annual and long-term trends in water-quality and biotic conditions in the reservoirs and their contributing watershed tributaries.

c) To provide data, interpretations from analysis of the data, and (or) conceptual or simulation models to support development and verification of predictive tools for use in managing the watersheds, reservoirs, and water supply.

d) To demonstrate that restorative and protective actions actually result in reduction or elimination of contaminants or other adverse impairments to the water supply.

e) To demonstrate that RWMA goals and Federal [SDWA and CWA 303(d)] and State standards are being met

in the watershed tributaries, the reservoirs, and the finished water supply.

On the basis of the preceding discussions on water-quality monitoring in both the watersheds and reservoirs, it is evident that the RTG has made considerable progress in achieving their monitoring objectives and enabling the RWMA partners to address their water-quality concerns. It also is evident that the RTG recognizes that to fully meet the RWMA goals and their objectives ultimately requires (a) the identification and implementation of restorative actions, and (b) that these restorative efforts in the reservoir watersheds could take years to reduce concentrations of nonpoint source contaminants (for example, sediment, nutrients, sodium, and chloride) in the tributaries and reservoirs to the point that their influence on watershed or reservoir water and biotic quality are no longer of concern. Furthermore, in the case of emerging water-quality concerns, such as DBPs, it is recognized that considerably more work needs to be done to identify the sources of organic carbon that lead to elevated DBP concentrations in finished waters in the City water-supply distribution system.

Along with these successes and pragmatic limitations in achieving RWMA goals and meeting RTG objectives, it is evident from the preceding discussions that the design of the monitoring program warrants modifications if continued progress is to be made. Possible improvements in the monitoring program can be described in relation to three broad areas: a) the monitoring design framework, b) the spatial and temporal resolution of water-quality assessments in the major tributaries and reservoirs, and c) the management and archival of data.

The improvements in these three areas incorporate suggested modifications to the monitoring program that were described and warranted on the basis of previous studies, but appear to not have been adopted, and (or) reflect inadequacies identified as a result of this retrospective review (table 12). The first two tabulated modifications are broad in scope—to modify monitoring to provide improved temporal and spatial coverage of water quality and biotic conditions associated with intra-seasonal, seasonal, and annual variations in climate, streamflow, and reservoir hydrodynamics, and to develop a comprehensive QAPP. Adoption of these modifications could help improve the ability to describe, model, compare, and address water-quality and biotic conditions associated with every long-term, emerging, and (or) regulatory (TMDL-related) water-quality concern. The remaining proposed modifications are narrow in scope, and focus on individual long-term and (or) emerging water-quality concerns (table 12). Adoption of one of these concern-specific modifications without the adoption of the two initial broad modifications could limit improvements in the ability to address that water-quality concern. Also, no attempt was made to prioritize one concern-specific modification over another as this would require a prioritization of RTG objectives and RWMA water-quality concerns, which is beyond the scope of this report.

Table 12.   Possible improvements identified in scientific or technical investigations that address one or more water-quality concerns of the Baltimore Reservoir Watershed Protection Program and required data from the long-term monitoring program.

[TMDL, total maximum daily load; chl-*a*, chlorophyll-*a*; DO, dissolved oxygen; TOC, total organic carbon]

Focus area(s) identified for improvements	Modification(s) to monitoring program framework	Report subsection(s) that further describe limitations addressed by modifications
Monitoring Program Data Requirements and Data Quality: To improve ability to readily and adequately understand the quality of the archived monitoring data and their suitability for interpretive purposes, and ability to readily understand the monitoring program design and data-quality requirements, which are both necessary in order to review, modify, or execute any major aspect of program. Current and sufficiently detailed descriptions of the data-quality requirements, quality-control and assurance procedures and data, and state of documentation of data in the long-term monitoring database were found to be inadequate. Long-standing data-quality issues do not appear to have been adequately addressed.	**Framework:** Develop a description of the monitoring program with specific quality-assurance and control plan that includes descriptions of the station networks, methods of data and sample collection, sample analyses, data-quality evaluation and verification, and data archival. Populate archival database using data with appropriate qualifying remarks and significant figures and related quality-control data.	**Review and Evaluation:** **Quality of the Monitoring Database and the Data Collected**

Table 12. Possible improvements identified in scientific or technical investigations that address one or more water-quality concerns of the Baltimore Reservoir Watershed Protection Program and required data from the long-term monitoring program.—Continued

[TMDL, total maximum daily load; chl-*a*, chlorophyll-*a*; DO, dissolved oxygen; TOC, total organic carbon]

Focus area(s) identified for improvements	Modification(s) to monitoring program framework	Report subsection(s) that further describe limitations addressed by modifications
General: To improve capability to assess effects of episodic or intra-seasonal climatic variations on watershed and reservoir hydrodynamics and associated water-quality and biotic conditions. Episodic-event and intra-seasonal pulses of contaminants appear to have an influence on most water-quality issues of concern, and will likely become increasingly important to consider given recent climate variability and forecasts for climate change. Addressing these limitations also will address most of the limitations in monitoring related to (a) describing and relating tributary water-quality and biotic conditions to watershed land-use activities, and management of those activities to reduce pollutant loads, and (b) relating tributary water-quality conditions to reservoir water-quality and biotic conditions either directly or through improving the data necessary for modeling these relations.	**Watershed Tributaries:** Collect high-flow (storm) samples (5-15 per year) and monthly fixed-time interval low-flow samples (12 per year) for water-quality and biotic conditions at seven continuous-discharge tributary monitoring sites in the reservoir watersheds; collect similar samples monthly (fixed-time interval, variable flow) at eight additional partial-discharge-record monitoring stations in the reservoir watersheds. Collect continuous depth-profile measurements for selected parameters at each monitoring site at onset of sampling. **Reservoirs:** Collect bimonthly fixed-time intervals and thermocline-dependent fixed-depth intervals samples for water-quality and biota at eight stations in three reservoirs during reservoir stratification; collect similar samples monthly (fixed-time interval) at these same stations during reservoir turnover. Collect continuous depth-profile data for selected water-quality and biotic parameters at six of the eight stations in the three reservoirs. **Hydrodynamics:** Obtain (a) continuous stream-discharge measurements for monitoring stations that include storm sampling; (b) partial-record stream-discharge measurements for monitoring stations that do not include storm sampling; (c) daily water levels, and volume of withdrawals, and, when applicable, releases, from each reservoir. Obtain the daily total amount and type of precipitation.	**Review and Evaluation: Modeling to Address Water-Quality Concerns and Monitoring to Address Individual Water-Quality Concerns, all concerns**

Table 12.   Possible improvements identified in scientific or technical investigations that address one or more water-quality concerns of the Baltimore Reservoir Watershed Protection Program and required data from the long-term monitoring program.—Continued

[TMDL, total maximum daily load; chl-*a*, chlorophyll-*a*; DO, dissolved oxygen; TOC, total organic carbon]

Focus area(s) identified for improvements	Modification(s) to monitoring program framework	Report subsection(s) that further describe limitations addressed by modifications
Eutrophication: To improve ability to quantitatively describe eutrophication in relation to mesotrophic and eutrophic phytoplankton conditions and water-quality (nutrient) and hydrodynamic (stratification) processes that affect phytoplankton production. Limited ability to relate short-term or long-term changes in phytoplankton taxa or abundance to reservoir water-quality conditions antecedent to their occurrence. Levels of available phosphorus and (or) nitrogen in the reservoir that lead to excessive algal blooms, or specific types of taxa blooms are unknown. Most of the annual load in available nutrients is transported during stormflows, which are insufficiently monitored to provide reasonably accurate estimates of storm loads. Whether or not there are in-lake sources of available nutrients promoting phytoplankton production has not been adequately assessed. Algal taxa and counts that occur in the winter and early spring are not adequately characterized, especially in Prettyboy Reservoir. Data limitations prevent accurate (a) comparisons among subbasins on basis of total yields, (b) estimation of tributary nutrient loads, (c) estimation of the available nutrients in the reservoirs and their influence on phytoplankton production, and (d) adequate characterization of reservoir hydrodynamics. All of the above also limit the ability to model the influence of tributary nutrient loads and possible in-lake sources on reservoir phytoplankton production, and the ability to identify what levels of available nutrients lead to excessive algal blooms. Sampling and data collection during stratification in actual epilimnetic layer possibly limited to only two depths. Insufficient monitoring frequency for chl-*a* and DO to address phosphorus TMDL endpoints for Loch Raven and Prettyboy Reservoirs.	**Framework:** Consider adoption of a broad descriptive model for seasonal phytoplankton production (taxa and abundance); evaluate model on the basis of monitoring data; define and describe stratification (epilimnion, metalimnion, and hypolimnion) on the basis of actual thermocline; describe mesotrophic and eutrophic conditions for each reservoir (a) on the basis of trophic indices determined for actual epilimnion and hypolimnion, (b) in relation to spatial differences within the reservoir, and (c) temporally—seasonally and as long-term trends; utilize watershed reservoir model to link nutrient loads from tributaries to reservoir trophic conditions. **Watersheds:** Collect nutrient (total, total dissolved, and dissolved phosphorus; total Kjeldahl, nitrate (plus nitrite), and ammonium nitrogen) samples during all sampling at seven continuous-discharge tributary monitoring sites in the three reservoir watersheds; collect similar samples and data during monthly sampling at all eight partial-record stations in the three reservoir watersheds. Obtain continuous depth-profile measurements for water temperature, DO, specific conductance, pH, chl-*a*, and turbidity during each sampling event at every station. Collect similar samples and data for all reservoir releases. **Reservoirs:** Collect nutrient (see Watersheds, above) and chemical oxygen demand samples during depth-profile sampling at all eight stations in three reservoirs; obtain continuous (daily) depth profile measurements for water temperature, DO (diurnal, during stratification), specific conductance, pH, chl-*a*, turbidity, and transparency at six of eight reservoir stations; obtain intermittent depth-profile measurements for these same parameters during sampling at remaining two reservoir stations. Sampling in actual epilimnetic layer should include three depths. Collect manganese and iron samples during collection of hypolimnetic samples at all eight stations.	**Review and Evaluation: Modeling to Address Water-Quality Concerns and Monitoring to Address Individual Water-Quality Concerns, Long-Term Concerns, Eutrophication**

Table 12. Possible improvements identified in scientific or technical investigations that address one or more water-quality concerns of the Baltimore Reservoir Watershed Protection Program and required data from the long-term monitoring program.—Continued

[TMDL, total maximum daily load; chl-*a*, chlorophyll-*a*; DO, dissolved oxygen; TOC, total organic carbon]

Focus area(s) identified for improvements	Modification(s) to monitoring program framework	Report subsection(s) that further describe limitations addressed by modifications
Sedimentation: To improve ability to quantify sediment transport in the watershed tributaries and its influence on reservoir water quality and storage capacity. From the mid-1980s to mid-1990s, 90 percent or more of the annual sediment load from six major subbasins was transported in storms. There has been a lack of adequate tributary storm data for suspended sediment in supply reservoir watersheds since 1995, and possibly to date. Lack of data limits (a) comparisons of reservoir subbasins as source or restored areas, and (b) estimation of annual loads, either directly or model-aided for all tributaries, and (c) determination of the effects of suspended sediment on reservoir water-quality and biotic conditions. Model simulations also are needed to help establish TMDLs for Liberty Reservoir and indicate compliance with existing TMDLs for sediment loads to Loch Raven Reservoir. No storm sampling has been conducted on any major tributary to Prettyboy Reservoir. Increases in turbidity at Prettyboy Reservoir indicate that establishment of a sediment TMDL could eventually be necessary.	**Framework:** Upgrade at least one tributary monitoring station on Prettyboy Reservoir watershed to act as a continuous-discharge monitoring site. **Watersheds:** Conduct suspended-sediment and turbidity-depth profiles during all sampling at seven continuous-discharge tributary monitoring sites in the three reservoir watersheds; collect similar samples and data during monthly sampling at eight partial-record stations in the three reservoir watersheds. **Reservoirs:** Conduct suspended-sediment sampling during all sampling at eight reservoir stations; obtain continuous (daily) profile monitoring for turbidity at six of these eight stations, and intermittent profile monitoring for turbidity during sampling at remaining two reservoir stations.	**Review and Evaluation: Modeling to Address Water-Quality Concerns and Monitoring to Address Individual Water-Quality Concerns, Long-Term Concerns, Sedimentation**

Table 12. Possible improvements identified in scientific or technical investigations that address one or more water-quality concerns of the Baltimore Reservoir Watershed Protection Program and required data from the long-term monitoring program.—Continued

[TMDL, total maximum daily load; chl-*a*, chlorophyll-*a*; DO, dissolved oxygen; TOC, total organic carbon]

Focus area(s) identified for improvements	Modification(s) to monitoring program framework	Report subsection(s) that further describe limitations addressed by modifications
Mercury in Higher Trophic Order Game Fish: To improve understanding of internal reservoir biotic and hydrodynamic processes that contribute to mercury methylation and biotic uptake. Sources of mercury for reservoirs were shown to likely be predominantly atmospheric and from regional sources beyond reservoir watershed boundaries. Enhanced mercury methylation and biological uptake, however, could occur during reservoir recovery following severe drought and reservoir drawdown in reservoir areas where terrestrial vegetation became established on exposed lake bed sediments. Current monitoring is insufficient to determine whether or not this occurs. Current lack of understanding of the factors that affect mercury methylation and biological uptake in Maryland lakes and reservoirs makes it difficult to determine without adequate monitoring whether or not enhanced methylation and uptake of mercury occurs during recovery in the Baltimore reservoirs.	**Framework:** Routinely sampling young of year trophic game fish in reservoir areas where bed sediments likely will become exposed and covered by terrestrial vegetation is necessary to provide a baseline before drought occurs. Given that drought conditions lead to the establishment of terrestrial vegetation, resampling during the year of recovery and year thereafter would provide the data needed to determine whether these conditions lead to enhanced mercury methylation and biological uptake in trophic game fish.	**Review and Evaluation: Monitoring to Address Individual Water-Quality Concerns, Long-Term Concerns, Sedimentation, Sediment Diagenesis, and Mobilization of Metals and Phosphorus**

Table 12. Possible improvements identified in scientific or technical investigations that address one or more water-quality concerns of the Baltimore Reservoir Watershed Protection Program and required data from the long-term monitoring program.—Continued

[TMDL, total maximum daily load; chl-*a*, chlorophyll-*a*; DO, dissolved oxygen; TOC, total organic carbon]

Focus area(s) identified for improvements	Modification(s) to monitoring program framework	Report subsection(s) that further describe limitations addressed by modifications
Disinfection By-Products (DBPs): To improve capability to identify sources of carbon that form DBPs. DBPs do not appear in raw water. Reducing the amount of TOC in raw intake waters at treatment plants by approximately one-third is insufficient in the prevention of DBP formation after chlorination in the distribution system at selected stations at concentrations that likely will exceed pending Federal drinking-water standards, particularly for total trihalomethanes. Current monitoring of only TOC appears insufficient to show that elevated levels of carbon that form DBPs are present. Sampling of TOC routinely has not been conducted on raw-intake water that subsequently is treated and then sampled for TOC and ultimately for TOC and DBPs in the distribution system.	**Framework:** The amount and type (brominated or chlorinated) of trihalomethanes formed is highly dependent on variations in (a) the types and levels of carbon that form DBPs, which may or may not be reflected by similar variations in the level of TOC, and (b) the treatment process, including the levels of initial chlorination, whether or not any DBPs that form during initial chlorination are reduced by coagulation and filtration, and whether or not finished water is initially stored, the length of time it is stored, and the manner of storage, whether or not stored water is re-chlorinated before it is placed in the distribution system, and the length of residence or traveltime the finished water travels in the distribution system. Minimally, sampling for TOC in raw water, sampling for TOC in treated water, and sampling for TOC and DBPs in the distribution system all must involve generally the same water if the use of TOC as a surrogate for DBP forming carbon, and, in particular trihalomethanes, is to be evaluated. Indirect evidence presented in this report indicates that possible sources of carbon that form trihalomethanes could routinely include phytoplankton (residues) in the reservoirs. An additional source of carbon could be terrestrial vegetation that grows on reservoir bed sediments exposed by excessive drawdowns during severe droughts that is then submerged during reservoir recovery.	**Review and Evaluation:** Monitoring to Address Individual Water-Quality Concerns, Emerging Concerns, Disinfection By-Products

Table 12. Possible improvements identified in scientific or technical investigations that address one or more water-quality concerns of the Baltimore Reservoir Watershed Protection Program and required data from the long-term monitoring program.—Continued

[TMDL, total maximum daily load; chl-*a*, chlorophyll-*a*; DO, dissolved oxygen; TOC, total organic carbon]

Focus area(s) identified for improvements	Modification(s) to monitoring program framework	Report subsection(s) that further describe limitations addressed by modifications
Sodium and Chloride: To improve ability to characterize and track changes and trends in sodium and chloride concentrations in watershed tributaries and reservoir waters. Historical data released from the monitoring study to date have not clearly defined the strength of correlations among conductance and the concentrations of chloride and sodium. Presumably this could be done for historical conditions. To determine the current strength of these correlations, however, monitoring data could be obtained for all three of these parameters as part of a monitoring program designed for that purpose. Current monitoring design is inadequate to provide early warning to water purveyors of the potential rise in sodium, and also inadequate to rapidly determine changes in sodium and chloride if management activities are introduced to reduce sodium chloride use as a deicing agent. If road deicing is no longer done with sodium chloride salt over the long term, there remains the larger question of whether sodium, chloride, or both will still continue to rise. Long-term data on chloride indicate it has steadily risen in concentration since the 1930s.	**Framework:** Select subbasin in each supply reservoir thought to contain smaller subbasins with high use of sodium-chloride salt as a deicing agent. Conduct baseline data collection for several years in smaller subbasin, at major subbasin, and in supply reservoirs at station closest to subbasin tributary. Implement management procedures to reduce salt use in smaller subbasins and continue monitoring to determine effect of reduced salt use. Ideally, subbasins selected lead to tributary inflows near reservoir station with continuous depth-profile monitoring. **Watersheds:** Collect sodium, chloride, and conductance samples as part of routine winter (November through March) monitoring in smaller and larger subbasin. **Reservoirs:** Collect sodium and chloride samples as part of all winter sampling conducted at reservoir station closest to tributary subbasin inflows. Include conductance in continuous depth-profile monitoring, or intermittent depth-profile monitoring, depending on reservoir station chosen.	**Review and Evaluation: Monitoring to Address Individual Water-Quality Concerns, Sodium and Chloride**

Modified Monitoring Framework

On the basis of this review, major modifications to the monitoring framework design for the Baltimore Reservoirs that could be considered to improve the ability of the RTG to address the water-quality concerns of the RWMA partners are as follows:

a) The adoption of a formal phytoplankton model for the reservoirs, such as the Plankton Ecology Group (PEG) model (Appendix D), which could be used to quantitatively describe intra-seasonal changes in taxa, their abundance, and succession in relation to variations in nutrient availability;

b) The collection of PEG-model compatible (total and available) nutrient (phosphorus and nitrogen) data in the reservoirs and watershed tributaries. These data could be used to improve load estimates, modeling of tributary loads and reservoir nutrient recycling, and relate changes in nutrient concentrations, loads, or ratios to temporal changes in phytoplankton production as described by the PEG model;

c) The collection of the appropriate hydrologic data to link hydrodynamic processes with water-quality, abiotic, and biotic processes. Collection of these hydrodynamic data would meet recommendations from modeling studies that indicated these data are necessary to accurately model, and to verify modeled reservoir watershed runoff, tributary flows, and reservoir hydrodynamics over the full flow regime of each watershed major subbasin tributary, and reservoir hydrodynamic response to releases or withdrawals and tributary flow regime, and

d) The use of statistical and modeling methods, in addition to graphical analysis, to provide scientifically defensible support to descriptions of changes in state, trends, or relations among watershed, tributary, and reservoir characteristics that involve the use of monitoring data given that the results of these analyses are used to promote action strategies that affect human activities in the reservoir watersheds.

A long-standing water-quality concern of the RWMA partners and therefore, area of investigation for the RTG, is reservoir eutrophication accompanied by major algal blooms. To address this issue directly, consideration could be given to the adoption of a formal model for phytoplankton production in the reservoirs. Such a model could provide a framework to not only describe taxa and their abundance, but intra-seasonal, seasonal, year to year, and multi-year changes in phytoplankton production and succession, and help to distinguish mesotrophic conditions from eutrophic conditions with excessive algal blooms in relation to variations in nutrient (phosphorus and nitrogen) availability. The PEG Model (**Appendix D**) by Sommer and others (1986) describes patterns in phytoplankton

succession that are similar to those described earlier in this report (see **Eutrophication**, *this report*). In addition, the PEG model takes into account that succession often is governed by variations in nitrogen, as well as phosphorus, availability, and other biotic processes.

Although the PEG model (or a similar phytoplankton model) could provide a useful reservoir framework for the Baltimore Reservoirs, the benefits from the use of this model described above require the collection of comparable and compatible nutrient data that can quantify either the total or available concentrations of nutrients in the reservoir-watershed tributaries and reservoirs. The comparable and compatible data include total Kjeldhal or organic nitrogen, in addition to available ammonia- and nitrate-plus-nitrite-nitrogen, and bioavailable phosphorus [DOP and (or) TDP] in addition to TP. Collection of these data also would enhance modeling capabilities to link water-quality conditions in the reservoir watershed tributaries to water quality and phytoplankton production and succession in the reservoirs (Interstate Commission on the Potomac River Basin, 2006).

Data obtained on variations in the concentrations and ratios of either total N to total P (or available forms of N to available forms of P) could be used to help explain observed PEG patterns in the seasonal succession of phytoplankton and (or) major bloom occurrence and dominance of selected taxa among phytoplankton communities in each of the reservoirs. This approach has proven to be useful as a diagnostic and management tool (Rast and Lee, 1978; Ryding and Rast, 1989; Levich, 1996; Bulgakov and Levich, 1999; Tõnno, 1999; Downing and others, 2001).

In addition, the above studies are highly relevant to concerns raised by the RTG that phosphorus is the limiting nutrient, and blue-green algal blooms create the greatest problems in the treatment of reservoir water and nuisance problems in treated water. For example, Rast and Lee (1978) and Ryding and Rast (1989) pointed out that the limiting nutrient can best be determined by measurements of available phosphorus (for example, DOP and (or) TDP) and nitrogen (for example, ammonia- plus nitrate plus nitrite nitrogen) during the period of maximum algal biomass. This has never been done in the Baltimore Reservoirs. Downing and others (2001) found that blue-green algal bloom abundance in 99 lakes correlated more strongly with variations in nitrogen and phosphorus, and standing algal biomass, than the simple single ratio of total N to total P. They also found that high standing algal biomass correlates with a high likelihood of blue-green algal dominance. The variations in available nutrients and in standing algal biomass with blue-green algal bloom occurrence have never been directly investigated for the Baltimore Reservoirs.

The collection of comparable types of nutrient data without the relevant and related hydrologic data could severely limit the use of the nutrient data. On the basis of reviews conducted on modeling studies (see **Modeling to Address Water-Quality Concerns**, *this report*), the availability of nitrogen in its various forms is governed by reservoir stratification. Available phosphorus also could be governed by reservoir stratification. Hence, the routine determinations of the thicknesses of the epilimnion, metalimnion, and hypolimnion are needed to accurately characterize nutrient availability for phytoplankton production. Reservoir water temperature is among the most influential and measureable factors that affect stratification. The elevation of reservoir water levels is among the most influential and measureable factors influencing reservoir temperature. Precipitation, tributary inflows, water withdrawals, and reservoir releases are among the most influential factors governing reservoir water levels. Thus, collection of all of the above climate and hydrologic data on a daily-to-weekly basis generally is considered a standard requirement to obtain the data necessary for reasonably accurate two-dimensional model simulations of the hydrodynamics and water-quality conditions of Mid-Atlantic Coast lakes and reservoirs using U.S. Army Corps of Engineers CE-QUAL-W2 (Giorgino and Bales, 1997; Bales and Giorgino, 1998; Sarver and Steiner, 1998; Bales and others, 2001; Galloway and Green, 2004, 2006 a,b).

For the RTG to be able to fully address RWMA concerns, it is equally important that the collection of biotic and water-quality data within each of the major reservoir watershed tributaries adequately reflect the range in tributary flows within a year and from year to year. This coverage could become increasingly critical in the future. Changes in climate and climatic variability (see **Effects of Climate**, *this report*) imply longer warmer and drier periods, punctuated by more intense storms. These changes combined with an ever-increasing population, and thus demand for water, could lead to reservoirs that are frequently drawn down to low levels, with recovery waters being of poor quality. This type of tributary flow coverage clearly has been a challenge for the Baltimore Reservoir monitoring program. Nevertheless, most of the water-quality concerns of the RWMA can be linked to excessive nutrient, sediment, salt, and other loads from the watershed tributaries. Without the ability to accurately assess the onsite effectiveness of BMPs to reduce loads, the most reliable means to assess whether these practices are effective is to measure downstream loads. On the basis of this review, the bulk of the annual loads to the Baltimore Reservoirs are transported during storm- or high flows.

Water-quality conditions associated with high flows have not been adequately covered by the Baltimore Reservoir monitoring program. To illustrate, sampling of stormflow in the Baltimore Reservoir watershed tributaries is generally conducted with automated samplers, or manually if necessary. Storm sampling involves the collection of water-quality samples at a gaged tributary within a single day. To assess the adequacy of this type of sampling, stream discharge from 2001–06 was summarized for each of the three major gaged tributary subbasins in each water supply reservoir. The summary, which includes the drought-recovery period from 2001–04, describes the proportion of each year in percent or number of days represented by the 75th percentile or higher streamflows (or generally the annual streamflow that reflects high flows), hereafter, referred to as high-flow days (fig. 32).

For example, for the period of record, and depending on the year, 25 percent of the annual flow at Morgan Run tributary in Liberty Reservoir watershed occurred during 4–11 percent (or 15 to 40 high-flow days) of the year. Monitoring records, however, indicate that for this entire period (4 years), water-quality data were only collected for seven storms, or approximately 7 high-flow days (table 7). In particular, during the driest year (2002), there were from 22–46 high flow days in 2002, depending on the tributary (fig. 32). Only one storm (1 high-flow day), however, was monitored at the three stormflow stations in Loch Raven Reservoir watershed, and no storms were monitored at the three stormflow stations in Liberty Reservoir watershed. However, monthly monitoring of dry-weather flows at all stations in 2002 could have included additional high-flow days because it was a very dry year. A different picture of high-flow coverage emerged during the wettest year of record (2004). There were from 14–36 days of very high flow, depending on the tributary (fig. 32). Only three storms (3 high-flow days) were monitored, however, at the stations in Loch Raven Reservoir watershed, and only one storm (1 high-flow day) was monitored at the three stations in Liberty Reservoir watershed. It is unlikely that monthly dry-weather flow monitoring included any additional high-flow days given the magnitude of high flows. Even if five storms (5 high-flow days) had been monitored during 2004 at each tributary station, it is unlikely that the full range in high-flow conditions was adequately represented at all six stations.

Adequate coverage of high flows within a year generally implies sampling is needed throughout the year and that consideration is given to the types, not just the number, of storms that occur. Typical storms that can occur over the Baltimore Reservoir watersheds are thunderstorms, nor'easters, and tropical storms (Mogil and Seaman, 2009). The origin and nature of these storms differ markedly in their spatial extent, duration, and the intensity and amount of precipitation, which can lead to marked differences in runoff response. They also are likely to occur at different times of the year.

To adequately reflect water-quality conditions associated with high and low flows, or the flow regime, within a given year, it is proposed that monitoring be conducted for both types of flows at sufficient frequency throughout the year at selected tributary stations in each reservoir watershed. It also is proposed that monitoring at the remaining traditional dry-weather flow stations continue to be conducted on a monthly fixed-time interval, regardless of flow condition. Details are provided below (see **Enhanced Spatial and Temporal Resolution of Water-Quality Assessments**, *this report*).

Sampling to adequately reflect the full range in flows within a given year, and from year to year, could markedly improve load estimates. It also could enhance the ability of the RTG to detect annual, seasonal, and discharge-related components in the water-quality and biotic variables of primary interest to the RTG. For example, Hirsch and others (2010) recently simultaneously identified long-term increasing or decreasing trends, seasonal variations, and major flow-related changes in biotic and water-quality conditions in nine large

[ft³/s, cubic feet per second]

Morgan Run near Louisville, Md.

| | | | | Annual streamflow characteristics | |
Water year	Mean daily flow (ft³/s)	Coefficient of variation (percent)	Total of mean daily flows (ft³/s)	Annual high-flow (75th-percentile flow or greater) characteristics
2001	20	85	7,315	25 percent of annual streamflow occurred in just 31 high-flow (39 ft³/s or higher) days, or during less than 10 percent of the year
2002	10	69	3,506	25 percent of annual streamflow occurred in just 39 high-flow (15 ft³/s or higher) days, or during less than 11 percent of the year
2003	50	113	18,417	25 percent of annual streamflow occurred in just 22 high-flow (109 ft³/s or higher) days, or during less than 6 percent of the year
2004	54	162	19,859	25 percent of annual streamflow occurred in just 15 high-flow (146 ft³/s or higher) days, or during less than 4 percent of the year
2005	34	128	12,589	25 percent of annual streamflow occurred in just 18 high-flow (84 ft³/s or higher) days, or during less than 5 percent of the year
2006	34	116	11,335	25 percent of annual streamflow occurred in just 24 high-flow (60 ft³/s or higher) days, or during less than 7 percent of the year

Selected streamflow characteristics for dry (2002) and wet (2004) water years

Reservoir/ watershed	U.S. Geological Survey streamgage[1]	Annual streamflow characteristics (ft³/s)				Number of days with high flows (75th-percentile flow or greater) (ft³/s)		Percentage of year with high flows (75th-percentile flow or greater)	
		Mean daily flow (2002)	Mean daily flow (2004)	Total of mean daily flows (2002)	Total of mean daily flows (2004)	2002	2004	2002	2004
Liberty Reservoir	Beaver Run near Finksburg, Md	5	25	1,971	9,038	31 (33 ft³/s or higher)	25 (208 ft³/s or higher)	9	7
	Morgan Run near Louisville, Md	10	54	3,506	19,859	39 (15 ft³/s or higher)	15 (146 ft³/s or higher)	11	4
	North Branch Patapsco River at Cedarhurst, Md	20	104	7,359	38,043	31 (33 ft³/s or higher)	25 (208 ft³/s or higher)	9	7
Loch Raven Reservoir	Beaverdam Run at Cockeysville, Md	13	42	4,567	15,507	20 (32 ft³/s or higher)	14 (146 ft³/s or higher)	5	4
	Western Run at Western Run, Md	20	103	7,427	37,515	46 (30 ft³/s or higher)	30 (175 ft³/s or higher)	12	8
	Little Falls at Blue Mount, Md	23	123	8,495	44,800	39 (34 ft³/s or higher)	31 (187 ft³/s or higher)	10	9
	Gunpowder Falls at Glencoe Road, Md[2]	91	343	33,381	125,301	44 (152 ft³/s or higher)	36 (564 ft³/s or higher)	12	10

[1] Streamgage names can differ from monitoring station names because of small differences in location

[2] Flows in 2002 notably reflect releases from Prettyboy Reservoir to resupply Loch Raven Reservoir

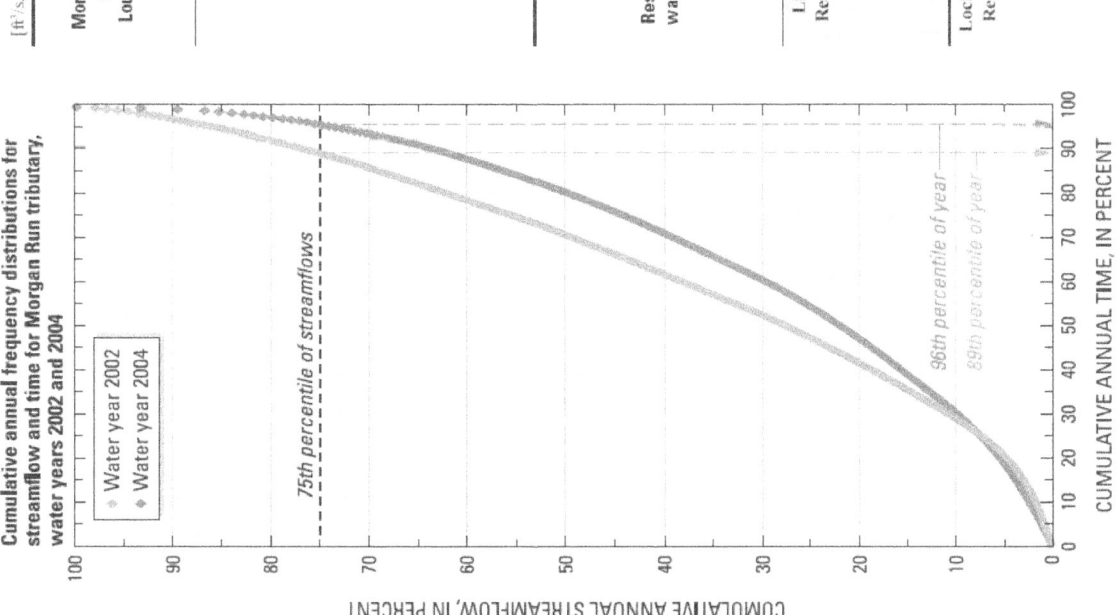

Figure 32. Cumulative annual distribution of streamflow and time with selected high-flow (75th percentile or greater) statistics for six streamgages associated with monitoring stations in Liberty and Loch Raven Reservoir watersheds from 2001–06, including driest (2002) and wettest (2004) years [Data from U.S. Geological Survey Maryland-Delaware-D.C. Water Science Center].

tributaries in the Chesapeake Bay using weighted regression analysis. Their work focused on many of the same water-quality parameters monitored in the Baltimore Reservoir system over approximately the same time period (1978–2008).

The work by Hirsch and others (2010) also illustrates the last major modification proposed for the design framework. The use of statistical analysis to characterize and compare changes in states, assess trends, describe seasonal variations, and (or) other relations among watershed, tributary, and reservoir characteristics could improve the RTG's ability and efficiency in understanding, identifying, and (or) managing the reservoir watersheds in relation to the RWMA concerns. Examples were presented in earlier sections of this report in which conclusions related to the above likely were drawn solely on the basis of appearance (graphical analyses) and not statistically verified. In selected cases, the lack of thorough statistical analyses early on led to the subsequent expenditure of considerable resources by contracted studies that possibly could have been avoided (for example, see **Eutrophication— Nutrients: Phosphorus Transport and Reservoir Recycling**, *this report*.) Additional examples that illustrate statistical analyses can be a useful tool to the RTG were also presented, including the following:

a) That there were significant declines in the concentrations of ammonia nitrogen at most tributary and reservoir stations throughout the 1980s–90s (see **Eutrophication—Nutrients: Nitrogen Transport and Reservoir Cycling**, *this report*), which indicates management activities to reduce the transport of nutrients, at least for this form of available nitrogen, do appear to be working;

b) That the lack of significant trends in TP at most tributary and reservoir stations likely reflected a step change and data outliers (see **Eutrophication—Nutrients: Phosphorus Transport and Reservoir Cycling**, *this report*), which indicated the need for improved documentation of changes in methods and archival of data, and

c) That the recent increases in sodium and chloride can be correlated with high road density in selected developed areas, and likely reflects the use of sodium-chloride salt as a deicing agent (see **Sodium and Chloride**, *this report*), which clearly helped the RTG identify a source that possibly could be managed to reduce sodium and chloride in drinking water.

In considering the proposed modifications, it also became apparent during this review that it could take considerable time, even with improvements in monitoring, before notable reductions in the frequency of eutrophic conditions and (or) major algal blooms can be statistically documented. For decades, elevated and varying concentrations of nutrients have

entered the reservoirs through tributary inflows. This has led to the development of a fairly robust, diverse, and abundant phytoplankton community in the reservoirs, to the extent that different algal taxa dominate in different years (see **Modeling to Address Water-Quality Concerns**, *this report*). Padisák (2004) noted that algal abundance and diversity provides for community resiliency; namely, a large standing and taxonomically diverse phytoplankton population has a finite but potentially large capability to absorb declines in levels of external resources, such as nutrients, without undergoing a substantial reduction in biomass.

Given that appreciable reductions in the frequency of eutrophic conditions and major algal blooms could take time, it is important that the monitoring program consider the above modifications to provide the type and quality of data most needed, and that the data be analyzed in a sound scientific manner. In the interim, adoption of the above modifications combined with improved statistical methods and models could enable the RTG to provide interim indications that management strategies are improving water-quality conditions in the tributary watersheds and water-quality and biotic conditions in the reservoirs. In addition, the information generated could help the RTG develop improved RWMA action strategies to reduce nutrient loads associated with the full range in hydrodynamic processes related to possible changes in climate, rather than chiefly in relation to dry-weather flows. Collectively, the interim results also could possibly encourage reservoir watershed residents to adopt management strategies long before the full benefits of those strategies clearly can be seen in the reservoir watershed tributaries, or ultimately in the reservoirs.

Enhanced Spatial and Temporal Resolution of Water-Quality Assessments

Modifications to improve the spatial and temporal resolution of water-quality assessments in the major tributaries and reservoirs of the Baltimore reservoir system can be described by use of a framework (table 13) similar in form, but modified in content, from the framework used for the existing monitoring design (**Appendix B**). The proposed framework reflects modifications to the existing monitoring in the reservoir watersheds and reservoirs which could improve the RTGs ability to address all RWMA water-quality concerns.

Watersheds

The proposed improvements to water-quality monitoring in the reservoir watersheds would utilize 15 of the tributary monitoring stations historically used in the monitoring network (table 13).[4] There would be six tributary stations in the

---

[4] Sampling of National Pollution Discharge Elimination System Sites or ponds (wastewater treatment plants or other point sources), when discharging, would continue as before but include any new water-quality parameters described for the reservoir watershed tributaries.

Table 13. Modified monitoring network for reservoir watershed tributaries and reservoirs in the Baltimore, Maryland drinking-water supply system.[1]

[---, not applicable; TBD, to be determined]

Reservoir/ watershed	Station location	Tributary station name	Station identifier[2]	Continuous profile monitoring	Fixed interval sampling[3]	Storm sampling[4]
Liberty	Upper reservoir	Liberty at Nicodemus/Deer Park Bridge	NPA0105	X	X	---
	Middle reservoir	Liberty at Oakland Road Point, near intake	NPA0067	---	X	---
	Lower reservoir	Liberty at gatehouse **or** Route 26 bridge (TBD)[2]	NPA0042 or NPA0059	X	X	---
	Tributary	Beaver Run at Hughes Road	BEA0015	X	X	X
	Tributary	Morgan Run at London Bridge Road	MOR0040	X	X	X
	Tributary	North Branch Patapsco River at Route 91	NPA0165	X	X	X
	Tributary	Middle Run at Louisville Road	MDE0026	---	X	---
	Tributary	Little Morgan Run at Bartholow Road	LMR0015	---	X	---
	Tributary (Pond)	Bonds Run at Hollingsworth Road	UZP0002		X	
Loch Raven	Upper reservoir	Loch Raven at Dulaney Valley Bridge **or** power lines (TBD)[2]	GUN0174 or GUN0190	X	X	---
	Middle reservoir	Loch Raven at Loch Raven Drive Bridge or between picnic and golf course areas	GUN0171	---	X	---
	Lower reservoir	Loch Raven at gatehouse	GUN0142	X	X	---
	Tributary	Beaver Dam Run at Beaver Run Lane	BEV0050	X	X	X
	Tributary	Gunpowder Falls at Glencoe Road	GUN0258	X	X	X
	Tributary	Western Run at Western Run Road	WGP0050	X	X	X
	Tributary	Dulaney Valley Branch at Loch Raven Drive	DVB0000	---	X	---
	Tributary	Gunpowder Falls at Falls Road	GUN0387	---	X	---
or	Tributary	Gunpowder Falls below Prettyboy Dam	GUN0398	---	X	---
	Tributary	Little Falls at Blue Mount Road	LIT0002	---	X	---
Prettyboy	Middle reservoir	Prettyboy at Beckleysville Road Bridge	GUN0437	X	X	---
	Lower reservoir	Prettyboy at gatehouse **or** 1,000 feet upstream of dam (TBD)[2]	GUN0399 or GUN401	X	X	---
	Tributary	Georges Run at Georges Creek Road	GOB0017	X, one station, TBD	X	X, one station, TBD
	Tributary	Graves Run at Gunpowder Road	GRG0013		X	
	Tributary	Gunpowder Falls at Gunpowder Road	GUN0476		X	

[1] Sampling also would be conducted at National Pollution Elimination Discharge Sites in each reservoir watershed if discharging, on a fixed interval (monthly) basis.

[2] For selected Liberty and Loch Raven Reservoir locations, the City of Baltimore and Baltimore and Carroll Counties, in consultation with the Reservoir Technical Group, need to determine which of the two stations would employ continuous profile monitoring. In a similar manner, the City and Counties would need to determine which tributary in Prettyboy Reservoir would have continuous monitoring and storm sampling.

[3] For reservoirs, bimonthly sampling during stratification, otherwise monthly. For tributaries with no storm sampling, conducted regardless of flow conditions. For tributaries with storm sampling, conducted at low-flow conditions.

[4] Conducted at high-flow conditions throughout the year, 5 to 15 samples per year.

Liberty Reservoir watershed, six tributary stations in the Loch Raven Reservoir watershed, and three tributary stations in the Prettyboy Reservoir watershed.

Sampling for high-flow conditions would be conducted at the three stations historically used for storm- and dry-weather flow sampling in each supply reservoir watershed, and in addition at one station in Prettyboy Reservoir watershed, which could be determined by the RTG. Upgrading one historical tributary station from dry-weather to full-flow monitoring in the Prettyboy Reservoir watershed is essential to address limitations noted in the attempts to describe water-quality and biotic changes in state, assess trends, estimate loads, or model tributary water quality (table 12). There simply are no long-term water-quality data for high flows on any Prettyboy Reservoir watershed tributary.

The seven tributary-monitoring stations targeted for high-flow sampling would be monitored for 5 to 15 storm events per station, selecting storms distributed throughout the year. Stormflow conditions at each station could be defined by the RTG on the basis of long-term existing streamflow records. Monthly fixed-time interval sampling also would be conducted during base flow at each of these seven stations. Base-flow conditions for each station could be defined by the RTG on the basis of historical flow records. In addition, continuous streamflow records would be maintained for these seven stations.

Monitoring at the remaining and historically dry-weather-flow stations in each reservoir watershed would be conducted on a monthly fixed-time interval, but regardless of flow conditions (table 13). Thus, monitoring at these stations over the course of each year would include base and high flows. In addition, partial streamflow records would be maintained for these stations.

To ensure results from stations within a given reservoir watershed are comparable, all stations within a given reservoir watershed that are sampled at high flows would be sampled for the same storm event. Minimally this would include all stations defined as storm-sampling stations in that reservoir watershed. It also could include additional stations sampled at monthly-fixed intervals depending on when the storm occurs during the month.

Water-quality samples collected at each and all monitoring stations for every sampling event would reflect representative samples for suspended sediment, or constituents for which concentrations could depend in part on suspended sediment, such as TP, total carbon, and total Kjeldahl nitrogen. Representative samples would require the collection of a depth- and width-integrated sample, unless it can be demonstrated that a single vertical depth-integrated sample or automated-sampler point sample is representative of the flow (Wilde, 2006). The latter sample-collection methods can be used if it is determined by a comparison of methods that they can provide a reasonably representative sample for the above constituents across the general range in streamflows at a station. No information was encountered in any report to indicate that such a comparative study has been done or if done was

ever repeated. Absent this information, any notable spatial bias in the quality of water flowing throughout the cross section could lead to a bias in the water-quality data collected from that cross section (Wilde, 2006). This bias could lead to misinterpretations when data from this monitoring station are compared to data from other monitoring stations, and could lead the RTGs to erroneous decisions, for example, in ranking subbasins for restorative action strategies.

To further help ensure that sample and data collection for every sampling event at each tributary monitoring station is representative, sampling at a station initially would include at least a mid-stream depth-profile (equidistant measurements throughout stream depth, minimum of three depths) for field measurements of water temperature, DO, conductance, pH, and turbidity using a portable multi-parameter water-quality monitoring system (table 13). These measurements could aid in the assessment of the uniformity of water-quality conditions throughout the depth of flow, and provide instantaneous values, for stream conditions associated with sampling.

Representative water samples collected and analyzed for every sampling event at a station could include samples for concentrations of not only suspended sediment, but sodium (minimally late October through early March), chloride, total and dissolved organic carbon, total, total-dissolved, and dissolved phosphorus, and total Kjeldahl- or organic-, nitrate (plus nitrite)-, and ammonia-nitrogen (table 14). Monitoring for dissolved carbon, total-dissolved and dissolved phosphorus, and total Kjeldahl (or organic) nitrogen addresses inadequacies in monitoring of total and dissolved forms of all three nutrients, which are described in earlier discussions in this report (table 12).

Monitoring of sodium in addition to chloride could be limited to winter months. Monitoring for sodium also could be limited if need be to selected subbasins within one supply reservoir watershed to establish baseline conditions before the implementation of any strategy designed to limit sodium-chloride use as road deicing agent. For example, one subbasin could be chosen and remain the control, and a second adjacent or nearby subbasin could be monitored for several years before and after being targeted for reductions in road salt use. Although strategies to reduce road-salt use could be implemented in all subbasins except for the control subbasin in this and the other two reservoir watersheds, the control and treated subbasins would provide the data needed to evaluate the rapidity and magnitude of change in sodium and chloride concentrations related to the reduction strategy.

To help ensure the quality of data and samples obtained from a tributary station, water-quality monitoring instrumentation for depth-profile measurements would be calibrated and operated according to manufacturer specifications. In addition, selected quality-assurance (QA) samples would be collected during each of two monthly fixed-interval sampling events per year (table 13), and at a randomly selected tributary station for each monthly fixed interval. Similar QA samples also would be collected for one storm event per year at one randomly selected tributary station. The QA samples would consist

Table 14. Modified data collection for hydrodynamic and water-quality constituents or characteristics at reservoir watershed tributary stations in the Baltimore, Maryland drinking-water supply system.

[X, representative sample; --, not collected; ft³/s, cubic feet per second; C, degrees Celsius; mg/L, milligrams per liter; µmhos/cm, micromhos per centimeter; TBD, to be determined (dependent upon sensor type)]

Hydrodynamic or water-quality constituent or characteristic	Sampling frequency[1]											
	March	April	May	June	July	August	September	October	November	December	January	February
Data collected at a tributary station during intermittent or storm sampling												
Streamflow, ft³/s	X	X	X	X	X	X	X	X	X	X	X	X
Temperature, air, °C	X	X	X	X	X	X	X	X	X	X	X	X
Temperature, water, °C	X	X	X	X	X	X	X	X	X	X	X	X
Dissolved oxygen, mg/L	X	X	X	X	X	X	X	X	X	X	X	X
Specific conductance, µmhos/cm	X	X	X	X	X	X	X	X	X	X	X	X
pH, standard units	X	X	X	X	X	X	X	X	X	X	X	X
Turbidity, TBD	X	X	X	X	X	X	X	X	X	X	X	X
Samples collected at a tributary station during intermittent or storm sampling[2]												
Suspended sediment, mg/L	X	X	X	X	X	X	X	X	X	X	X	X
Chloride, mg/L	X	X	X	X	X	X	X	X	X	X	X	X
Sodium, mg/L	X	--	--	--	--	--	--	--	X	X	X	X
Total organic carbon, mg/L	X	X	X	X	X	X	X	X	X	X	X	X
Dissolved organic carbon, mg/L												
Total phosphorus, mg/L	X	X	X	X	X	X	X	X	X	X	X	X
Total dissolved phosphorus, mg/L	X	X	X	X	X	X	X	X	X	X	X	X
Dissolved phosphorus, mg/L	X	X	X	X	X	X	X	X	X	X	X	X
Total organic or Kjeldahl nitrogen, mg/L	X	X	X	X	X	X	X	X	X	X	X	X
Nitrate-nitrogen, mg/L	X	X	X	X	X	X	X	X	X	X	X	X
Ammonia nitrogen, mg/L	X	X	X	X	X	X	X	X	X	X	X	X
Quality-assurance samples obtained during intermittent sampling[3]	--	..........X..........		--	--	--	--	--	............X............		--	--
Quality-assurance samples obtained during storm sampling for storm during designated period[4]	--	--	--	--	..........X..........		--	--	--	--	--	--

[1] Data and samples collected at National Pollution Discharge Elimination System sites and during supply reservoir releases and should include data and samples listed here.

[2] Storm sampling would be conducted for up to 12 (monthly) storm events per year at stations on three selected tributaries in Liberty Reservoir, and on one selected tributary in Prettyboy Reservoir watershed. If storm samples are collected using an automated sampler at a station, a visit would need to be made during the rising limb of storm runoff to obtain data normally collected during on-site sampling. If this is not possible, the following data normally collected on site—specific conductance, pH, and turbidity—would need to be obtained from measurements made on samples within 24 hours of collection.

[3] Obtained at one randomly selected station within each indicated 2-month period. Samples for suspended sediment would be collected; data compared to turbidity data. Field blanks and replicate water samples are collected for all constituents that will be obtained from the laboratory analyses of the collected water-quality samples. Include blind standard reference samples for as many of the constituents as possible that will be obtained from the laboratory analyses of the collected water-quality samples.

[4] Obtained at one randomly selected station within the 3-month period. Samples for suspended sediment would be collected; data compared to turbidity data. Field blanks and replicate water samples are collected for all constituents that will be obtained from the laboratory analyses of the collected water-quality samples. If automated system used for sample collection, blank is collected from system in field before storm; replicate samples automatically would be collected during storm.

of field blanks and replicate samples for all water-quality constituents determined by laboratory analysis. Field blanks would be collected on site before tributary sampling, with the sampling equipment used for tributary sampling, and for all constituents for which water-quality samples are sent for analyses. Replicate tributary (split or sequential) samples would be submitted for all constituents in water-quality samples sent for analyses.

In addition, and to the extent they are available at the appropriate concentrations, standard reference samples would be obtained and submitted as blind samples to the laboratories analyzing water-quality samples. These reference samples would be analyzed for concentrations of sodium, chloride, total and dissolved organic carbon, total, total-dissolved, and dissolved phosphorus, and total Kjeldahl- or organic-, nitrate (plus nitrite)-, and ammonia-nitrogen.

Reservoirs

In the modified network, there would be eight reservoir monitoring locations—the upper, middle, and lower parts of Liberty and Loch Raven Reservoirs, and the middle and lower parts of Prettyboy Reservoir (fig. 3, table 13). Monitoring locations in the lower part of each reservoir at or near each of the three reservoir gatehouses, in the upper parts of Liberty and Loch Raven Reservoirs, and in the middle part of Prettyboy Reservoir, would be equipped with continuous multi-parameter water-quality monitoring systems. Each of these systems would obtain daily depth-profile logs (minimum 5-ft intervals to 30 ft below water surface, and to-be-determined intervals below 30 ft) during daytime and (possibly nocturnal, for DO) during periods of stratification for seven parameters (table 15)—water temperature (°C or degrees Celsius), DO concentration (mg/L), specific conductance (microsiemens per centimeter), pH (standard units), chl-*a* concentration (fluorescence units), turbidity (to be determined, units dependent upon sonde probe type), and depth of measurement (ft) via an array of sensors automatically raised and lowered through the water column. The remaining two reservoir-monitoring locations would be in the middle parts of Liberty and Loch Raven Reservoirs (fig. 3). At these stations, depth profiles similar to those described above would be obtained bimonthly from March through October, and monthly from November through February, during the day. Transparency (Secchi-disc depth to extinction) would be measured at all stations during routine water-quality sampling.

Intermittent sampling would be conducted at all eight reservoir-monitoring locations (table 13). During periods of thermal stratification (March through October, table 15), samples would be collected twice monthly at each station at sufficient selected fixed depths similar to those used historically to characterize water quality associated with epilimnion, thermocline (metalimnion), and hypolimnion; during periods of thermal turnover (November through February), samples would be collected monthly at each station at similar depths to those used under the existing design with one exception.

During the period of general stratification, an additional sample would be collected at the 15-ft depth to ensure at least three samples are collected in the epilimnetic layer, which the Interstate Commission on the Potomac River Basin (2006) indicated could be shallower than 20–30 ft in thickness.

Samples collected at each depth would be analyzed for a variety of constituents (table 15), including concentrations of sodium (late October through early March only), algal taxa and counts (mid-March through October only), and concentrations of total and dissolved organic carbon, total, total-dissolved, and dissolved phosphorus, total Kjeldahl- or organic-, nitrate (plus nitrite)-, and ammonia-nitrogen, and chemical oxygen demand. In the hypolimnion, an additional sample would be collected for total manganese and total iron.

Among the proposed modifications to reservoir monitoring, the use of continuous multi-probe depth profilers could represent the most significant change in monitoring technology. Continuous-monitoring data are used elsewhere by reservoir and watershed managers. When used on a daily basis, they can provide the short-term water-quality data for key parameters necessary to relate watershed tributary-reservoir hydrodynamics and the associated water-quality and biotic conditions in response to major storm events, and seasonal and annual variations in climate, including drought and recovery conditions (Rasmussen and McAllister, 2005; Rowland and others, 2006).

Continuous water-quality monitoring instrumentation would be calibrated and operated according to manufacturer specifications. Well-developed guidelines and procedures address the installation, operation, and routine maintenance of continuous water-quality monitors, and describe how to address and verify the quality of the resultant data, which can be excellent with proper routine (typically bimonthly) field maintenance (Rowland and others, 2006; Wagner and others, 2006).

Selected QA samples would be collected during each of two fixed time periods (table 15), and at a randomly selected reservoir station for each fixed time interval. The QA samples would consist of field blanks and replicate samples for all water-quality constituents determined by laboratory analysis. Additional water-quality samples would be obtained for comparative analysis between sonde measurements for chl-*a* (fluorescence units) and chl-*a* analysis (in μg/L), and possibly for sonde turbidity measurements (dependent upon probe type used). In addition, standard reference samples to the extent they are available would be obtained and submitted as blind samples for the analyses of as many water-quality constituents as possible, including concentrations of sodium, chloride, total-, total dissolved-, and dissolved phosphorus, total and dissolved organic carbon, and total Kjeldahl- or organic-, nitrate (plus nitrite)-, and ammonia-nitrogen.

The improved framework for the collection of water-quality data needs to provide the hydrodynamic data necessary to relate watershed-reservoir hydrodynamic and water-quality conditions with reservoir hydrodynamic, water-quality, and biotic conditions (table 15). Thus, each reservoir would be

Table 15. Modified data collection for climatic, hydrodynamic, and water-quality constituents or characteristics at reservoir locations in the Baltimore, Maryland, drinking-water supply system

[D, daily; TDAP(6), twice daily automated profile measurement at six locations; DAP(6), daily automated profile measurement at six locations; TMMP(2), twice monthly manual profile measurement at two locations; MMP(2), manual monthly profile measurement at two locations; SDP, sampling depth profile; SB, sampling bimonthly; water samples collected between 3 feet below surface to 3 feet above bottom; S, sampling monthly; water sample(s) collected between 3 feet below surface to 3 feet above bottom; X, quality-assurance sample; ---, not sampled; in., inches; ft, feet Mgal/d, million gallons per day; C, degrees Celsius; mg/L, milligrams per liter; µmhos/cm, micromhos per centimeter; TBD, to be determined (dependent upon sensor type); #/mL, number per milliliter]

Location/constituent or characteristic	Sampling method and frequency											
	Stratified reservoir								Non-stratified reservoir			
	March	April	May	June	July	August	September	October	November	December	January	February
At reservoir gatehouse or outlet[1]												
Precipitation, total (in.) and form												
Water levels, ft												
Water withdrawals and releases, Mgal/d	D	D	D	D	D	D	D	D	D	D	D	D
Reservoir surface elevation, ft												
Temperature, air, C												
At reservoir monitoring location, and with multiprobe sonde or Secchi disc[2]												
Depths of measurement profile, ft												
Temperature, water, C												
Dissolved oxygen, mg/L	TDAP(6) or TMMP(2)	TDAP(6) or TMMP(2)	TDAP(6) or TMMP(2)	TDAP(6) or TMMP(2)	TDAP(6) or TMMP(2)	TDAP(6) or TMMP(2)	TDAP(6) or TMMP(2)	TDAP(6) or TMMP(2)	DAP(6) or MMP(2)	DAP(6) or MMP(2)	DAP(6) or MMP(2)	DAP(6) or MMP(2)
Specific conductance, µmhos/cm												
pH, standard units												
Chlorophyll-a, fluorescence												
Turbidity, TBD												
Transparency (Secchi disc), ft												

Table 15. Modified data collection for climatic, hydrodynamic, and water-quality constituents or characteristics at reservoir locations in the Baltimore, Maryland, drinking-water supply system.—Continued

[D, daily; TDAP(6), twice daily automated profile measurement at six locations; DAP(6), daily automated profile measurement at six locations; MMP(2), manual monthly profile measurement at two locations; TMMP(2), twice monthly manual profile measurement at two locations; SDP, sampling depth profile; SB, sampling bimonthly, water samples collected between 3 feet below surface to 3 feet above bottom; S, sampling monthly, water sample(s) collected between 3 feet below surface to 3 feet above bottom; X, quality-assurance sample; --, not sampled; in., inches; ft, feet Mgal/d, million gallons per day; C, degrees Celsius; mg/L, milligrams per liter; μmhos/cm, micromhos per centimeter; TBD, to be determined (dependent upon sensor type); #/mL, number per milliliter]

Location/constituent or characteristic	March	April	May	June	July	August	September	October	November	December	January	February
	\<-- Stratified reservoir --\>								\<-- Non-stratified reservoir --\>			
At reservoir monitoring station, and with discrete-depth sampler												
Depths of sampling profile, ft	SDP SDP	SDP SDP	SDP SDP	SDP SDP	SDP SDP	SDP SDP	SDP SDP	SDP SDP	SDP	SDP	DP	DP
Sodium, mg/L	S	--	--	--	--	--	--	--	S	S	S	S
Total algal count, #/mL	SB SB	SB SB	SB SB	SB SB	SB SB	SB SB	SB SB	SB SB	--	--	--	--
Algal taxa identification	SB SB	SB SB	SB SB	SB SB	SB SB	SB SB	SB SB	SB SB	--	--	--	--
Chloride, mg/L												
Dissolved organic carbon, mg/L												
Total organic carbon, mg/L												
Total phosphorus, mg/L												
Total dissolved phosphorus or orthophosphate phosphorus, mg/L	SB SB	SB SB	SB SB	SB SB	SB SB	SB SB	SB SB	SB SB	S	S	S	S
Total organic nitrogen, mg/L												
Nitrate-nitrogen, mg/L												
Ammonia nitrogen, mg/L												
Chemical oxygen demand, mg/L												
Hypolimnion, Manganese, mg/L												
Hypolimnion, Iron, mg/L												
Quality-assurance samples	........X[3]				........X[3]				........X[3]			X

[1] Water purveyors collect a variety of water-quality samples at or near the Loch Raven and Liberty reservoir intakes. To the extent that the reservoir monitoring station at or near the gatehouse can provide suitable data from profiles and interval sampling, their sampling requirements can be reduced. Conversely, if water purveyors can provide suitable data from their routine sampling, interval sampling at the lower end of each reservoir can be reduced.

[2] Historical monitoring in reservoirs has included dissolved solids (mg/L), and hardness and alkalinity (mg/L, as calcium carbonate), which could be included in interval sampling, but they have not appeared to be critical assessment parameters in relation to water-quality issues of concern.

[3] Quality-assurance samples are collected within each designated period, at one randomly selected station. Chlorophyll-a and suspended-sediment samples would be collected; data would be compared to chlorophyll-a and turbidity measured with multi-parameter sonde. Field blanks and replicate water samples are collected for all constituents that will be obtained from the laboratory analyses of the collected water-quality samples. For one randomly selected period, include blind standard reference samples for as many of the constituents as possible that will be obtained from the laboratory analyses of the collected water-quality samples.

characterized with respect to daily water levels, releases, and, in the case of the drinking-water reservoirs, withdrawals.

Each reservoir system also would be characterized with respect to mean daily air temperature and precipitation, with the latter including daily totals and form (rain or snow). Depending on the reservoir watershed, climate stations could be installed within the reservoir property controlled by the City, or within reservoir watershed property controlled by either Carroll or Baltimore Counties. Obtaining these data in proximity to the reservoirs and reservoir watersheds would improve modeling capabilities (see **Modeling to Address Water-Quality Concerns**, *this report*).

Susquehanna River

If the ongoing review of the City firming program results in an increased use of the Susquehanna River as a primary source of drinking water, and the SRBC agrees to this increase in use, then the RTG and City would need to determine the implications of this decision with respect to its long-term monitoring program. Use of Susquehanna River water as a drinking-water source will complicate the identification and management of source waters that lead to water treatment and finished water impairments. For example, concentrations of DBPs in finished waters in the City water-distribution system from the Montebello treatment facility could reflect treatment of waters obtained from Loch Raven Reservoir, the Susquehanna River, or both sources. Minimally, the RTG needs to ensure that periods of Susquehanna River water use are documented and archived for easy reference.

Enhanced Documentation of Data and Quality of Data Collection

Regardless of whether the monitoring network is modified as a consequence of this review, the RWMA requires that long-term (multi-decadal) monitoring be performed to obtain the data necessary to address its goals and water-quality concerns. Inherent in this effort is the need to be able to access data from the entire period during which they have been collected. It also is critical that the data are of known quality.

Invariably, field and laboratory personnel as well as the methods used to collect data and samples in the field or analyze samples in the laboratory change with time. To ensure these changes do not adversely affect the ability to obtain and use these data, and that the quality of the data being used is known or can be determined over an indefinite period of time, implies that data obtained in the field or from a laboratory be stored in a well-maintained and supported database with the appropriate qualifications (units, significant figures, and remarks). It also generally implies that the quality of the data entered be determined through the independent collection of QA data in the field and laboratory, and that these QA data be stored in a similar fashion, ideally either in the same database or at least a similarly supported companion database. These

operational objectives generally are successfully completed if they are guided by a single comprehensive QAPP.

On the basis of a review of available (2007) documents on data-collection methods, QAC plans, and the database, it is apparent that a single comprehensive QAPP would be of great benefit to the RWMA program. For this QAPP to be effective and eliminate the shortcomings identified in this review, it would need to include the following elements:

a) A description of the field and laboratory procedures and methods and organizations responsible for the collection of data and samples for the monitoring program.

b) Definitions of the long-term data-quality requirements for the monitoring program, including the identification of the appropriate reporting level, precision, and bias required (determined suitable) for each hydrodynamic, water-quality, or biotic parameter stored in the database.

c) A description of the required quality-assurance and control measurements (QAC data), the agencies responsible for their collection and entry, and the manner in which they routinely will be obtained and digitally stored along with the corresponding hydrodynamic, water-quality, or biotic parameters.

d) A description of the procedures for the routine (annual or biannual) summary, review, analysis, and evaluation of the QAC data, the agencies responsible for conducting these activities, and the documentation of findings, to determine if data requirements are being met.

e) A description of procedures for the systematic transfer or modification of field or laboratory methods, and required documentation thereof, and the responsible agencies for overseeing this activity given that data-quality requirements and (or) methods invariably change.

To be most effective, this QAPP would be developed and implemented by the RTG partners, and include the DPW and Baltimore County field operations and the DPW treatment facility staff and contract laboratory staff that analyze long-term monitoring samples.

Examples of comprehensive QAPPs, and the parts of these QAPPs that routinely undergo periodic change, for other long-term monitoring programs can be found in the following resources:

a) Quality Management Plans: U.S Environmental Protection Agency (2008) National Risk Management Research Laboratory's (NRMRL) Quality Management Plan (QMP);

b) Quality Assurance Program Plans: California Water Resources Control Board (2008) Surface Water Ambient Monitoring Program (SWAMP), with guidance from Keith (2006); or the Chesapeake Bay Program

(2008) non-tidal monitoring quality plan, which is modeled after the U.S. Environmental Protection Agency (2001) quality assurance program plan for measurement projects; and

c) Environmental Sampling Protocols Documentation: The U.S. Geological Survey National Field Manual for Water-Quality Sampling (U.S. Geological Survey, variously dated).

In addition, examples of the types of quality-control samples that can be collected as part of routine monitoring were described for the watershed tributaries (table 14) and reservoirs (table 15). The blind standard-reference samples referred to in these examples are available from inter-laboratory verification programs and commercial vendors. Two examples of regional and national programs that provide standard reference samples for water-quality parameters of interest to the RWMA partners are (a) the Split and Blind Sample Program of the Chesapeake Bay Program (Chesapeake Bay Program, 2010), and (b) the Inter-Laboratory Comparison Program of the USGS Branch of Quality Systems (Glodt and Pirkey, 1998).

Benefits of an Enhanced Water-Quality Monitoring Program

The proposed improvements to the long-term monitoring network reflect the combined recommendations or address the notable limitations described in the preceding discussions on the use of monitoring data to address watershed and reservoir water-quality issues of concern, and the monitoring-related objectives of the RTG. The improved monitoring design can provide the RTG with the type and quality of data needed to improve upon spatial and temporal assessments under potentially rapid changes to long-term changes in hydrodynamic and associated water-quality and biotic conditions. In particular, they would help the RTG to:

a) Describe states, changes, and trends in the concentrations and loads for the water-quality parameters used to address the water-quality concerns in the reservoir watersheds;

b) Describe states, changes, and trends in the concentrations of water-quality parameters used to assess, or compute indices to assess, water-quality and biotic conditions that reflect water-quality and biotic concerns in the reservoirs;

c) Describe states, changes and trends in the concentrations of water-quality and biotic parameters used by City water purveyors to identify reservoir intake(s) for raw-water withdrawals for water treatment, and (or) to reduce or eliminate potential impairments in finished drinking water;

d) Describe modeled states, changes, or trends in the concentrations or loads of water-quality and biotic parameters required to develop (Liberty Reservoir and watershed) or improve upon (Loch Raven and Prettyboy Reservoirs and watersheds) simulations of hydrodynamics and associated water-quality conditions in the reservoirs;

e) Quantify states, changes, and trends in concentrations and loads associated with water-quality and biotic parameters used to address regulatory requirements, including TMDLs, or water-quality parameters used as TMDL endpoints in the reservoirs;

f) Assess water-quality conditions associated with emerging water-quality issues of concern—concentrations and sources of DBPs and of sodium and chloride;

g) Improve understanding of the effects of climate (variability and change) on water-quality conditions in the reservoir tributaries, and water-quality and biotic conditions in the reservoirs associated with each long-term and emerging water-quality concern; and

h) Improve the reservoir-monitoring framework to provide continuous profile data possibly in real time, and at a higher frequency if needed, for water purveyors to provide advance warning of changes in the quality of reservoir water in the upper part of each water-supply reservoir that could imply eventual changes in water quality in the lower parts of each reservoir near the intakes used to withdraw water for drinking-water supplies.

Summary

The City of Baltimore, Maryland, and parts of five surrounding counties obtain water from Loch Raven and Liberty Reservoirs. A third reservoir, Prettyboy, is used to resupply Loch Raven Reservoir. Management of the watershed conditions in all three reservoirs is a shared responsibility among City, County, and State representatives under the Baltimore Reservoir Watershed Management Agreement (RWMA), which is guided by voluntary agreements and action strategies. Management of the reservoirs is chiefly the responsibility of the City. Withdrawals for supplies vary markedly in response to seasonal demands, and routinely aim to withdraw only the best quality of water for supplies to reduce treatment costs and complaints about the quality of treated water. However, they must be managed to reduce major drawdowns in all three reservoirs. Reservoir recovery, particularly for Liberty Reservoir, can take months following major withdrawals and drawdowns, and major drawdowns possibly can result in reductions in the quality of reservoir water during recoveries.

A major purpose of these RWMA agreements is to address long-term and emerging water-quality concerns. The 2005 RWMA action strategy called for continued comprehensive water-quality monitoring in the reservoirs and selected watershed tributaries related to these concerns, and indicated that monitoring could be revised as needed based on an evaluation of: a) its ability to detect changes in state or trends, b) its suitability in relation to the areal extent and adequacy of monitoring to provide the type and quality of needed data, and c) its effectiveness in helping to understand and manage the reservoir watersheds and reservoirs to address water-quality concerns. To aid in this effort, the U.S. Geological Survey conducted a review of the effectiveness of the monitoring program from its inception in 1982 through 2007 in relation to the above criteria.

Long-Term and Emerging Water-Quality Concerns

The long-term water-quality concerns of the RWMA partners include eutrophication and sedimentation in the reservoirs, and elevated concentrations of (a) nutrients (nitrogen and phosphorus) transported from the major tributaries to the reservoirs, (b) iron and manganese released from reservoir bed sediments during periods of deep-water anoxia, (c) mercury in higher trophic order game fish in the reservoirs, and (d) bacteria in selected reservoir watershed tributaries. Emerging concerns include elevated concentrations of sodium, chloride, and disinfection by-products in the drinking water from both supply reservoirs. Climate variability and climate change also could be emerging concerns. The inherent variability in climate could affect within-year and year-to-year variations in water quality in the tributaries and reservoirs. Projected changes in climate in the region include an increased frequency in drought conditions punctuated by an increased frequency in intense storms.

Eutrophic conditions that lead to major algal blooms are presumed to be caused by elevated nutrient (nitrogen and phosphorus) concentrations and loads from the major tributaries to the reservoirs. Elevated concentrations of iron and manganese are released from bed sediments under hypoxic reservoir conditions, which often follow the die-off and decomposition of major algal blooms. Sedimentation in the reservoirs largely is caused by elevated suspended-sediment concentrations and loads from the reservoir watersheds. Major sediment loads reflect overland as well as in-stream bed and bank erosion. Sediment transported to the reservoirs reduces reservoir storage capacity, increases in-lake turbidity, and is the major source of iron and manganese as well as a potential source of available phosphorus to the reservoirs. Elevated concentrations of mercury, which occur in higher trophic order game fish in the reservoirs, likely result from regional sources that are presumed to be beyond the control of the RWMA partners. Elevated concentrations of bacteria also occur in selected tributaries. As of 2007, their occurrence and sources were being investigated.

Emerging water-quality concerns (primarily related to drinking water) include elevated concentrations of disinfection by-products measured at selected times and monitoring stations in the water-supply distribution systems of both supply reservoirs. Concentrations of disinfection by-products could exceed Federal drinking-water standards upon implementation of a recent rule change. The source(s) of the carbon compounds that form these by-products, however, are largely unknown. Elevated concentrations of sodium, presumably from increased use of sodium-chloride salt as a deicing agent, also recently have approached or exceeded the Federal standard for individuals on low-sodium diets in drinking water from both supply reservoirs. Although not formally considered by the RWMA partners as an emerging concern, climate variability and climate change were considered in this review as emerging concerns. Climate variability in the Mid-Atlantic region already notably affects within-year and year-to-year variations in the quantity and quality of tributary and reservoir waters. Climate change for the Mid-Atlantic is projected to result in an increase in the frequency and length of hot dry-weather periods and in an increase in the number of intense storms. Historical patterns indicate the quality of waters in the reservoirs can decline during droughts that result in major drawdowns and (or) following the recovery of the reservoirs from such drought conditions.

Description of Water-Quality Monitoring

The chief purpose for monitoring in the Baltimore Reservoir System has been to provide data to describe, assess, and relate changes in the state of water-quality and biotic conditions in the major reservoir watershed tributaries and reservoirs associated with long-term and emerging RWMA water-quality concerns. In addition, monitoring has been conducted to support the development of models that could help the RWMA partners relate watershed and tributary conditions to reservoir conditions, in order to better identify either effective watershed management activities, or better assess the effectiveness of activities already performed, to address their water-quality concerns.

Since its inception in the early 1980s, the long-term core monitoring network for the Baltimore Reservoir System generally has consisted of approximately 21 nonpoint-source (tributary or pond) and point-source water-quality monitoring stations in the reservoir watersheds, and 12 in-lake monitoring stations in the three reservoirs. As of 2007, there were few changes in the network. The most notable documented changes in monitoring include a decline in the frequency of storm sampling in 1995, and the discontinuance of monitoring for bioavailable (dissolved phosphate) phosphorus in 1993. Undocumented or never fully explained changes in monitoring include the manner in which storm samples were collected in the mid-1980s, and the manner in which phosphorus samples were collected in the mid-1990s.

Review and Evaluation of Monitoring

The review and evaluation of the monitoring program included (a) a broad review of the quality of the monitoring database and dated collected from the inception of data collection in 1982 through 2007, (b) a review of the monitoring program and data in relation to their ability to support modeling to address water-quality concerns, and (c) a review of the monitoring program and data to support analyses that address each individual long-term and major water-quality concern.

Quality of Monitoring Database and Data

During the course of this review, concerns were encountered by RWMA partners or contractors about the quality of monitoring data. At least four investigative studies actually published those concerns as part of their investigative work. A broad review of the related database documentation as of 2007 revealed no single quality assurance program and planning (QAPP) document was readily available that described in sufficient detail the current reservoir monitoring program from data collection to archival. No document was found that describes the data-quality requirements for the monitoring program, or indicates that routine reviews with summaries are conducted by the Reservoir Technical Group (RTG) on the quality of data being collected by the monitoring program. It also was apparent that the quality-assurance and control plans for City water-treatment facility laboratories, which analyze most monitoring samples, cannot be expected nor used to meet the quality-assurance and control requirements of a long-term monitoring program. In addition, responsibilities for data collection and analysis for the Baltimore Reservoir monitoring program have been distributed among different City staff, City partners, or private contractors, leading to difficult data-quality issues that can arise long after the data have been collected.

Monitoring to Support Modeling to Address Water-Quality Concerns

Monitoring data were reviewed in relation to their ability to support the development of models that could help identify effective watershed management activities, or assess the effectiveness of activities already undertaken. Such modeling is viewed by the RTG, as an effective means to help address existing and future RWMA partner water-quality concerns. To date, model development has been a challenge because of a lack of required data or limitations in the availability of required data. A review of the most recent and comprehensive modeling effort was conducted as part of this study. A computer simulation model of the watersheds draining into Prettyboy and Loch Raven Reservoirs (Hydrological Simulation Program—Fortran, HSPF) coupled with a two-dimensional lake computer simulation model (U.S. Army Corps of Engineers CE-QUAL-W2) was used to simulate the hydrodynamics and water quality of the reservoirs. Use of this coupled model was adequate to link nutrient loads, specifically phosphorus loads, from the Loch Raven and Prettyboy Reservoir watersheds to algal biomass concentrations (represented by chlorophyll-a or chl-a concentrations) in the reservoirs, and to calibrate the relation between autochthonous and allochthonous organic matter and dissolved oxygen (DO) concentrations in the hypolimnetic layer. The models could be improved, however, with the addition of the appropriate data. Limitations in data noted by the modelers were considerable. In brief, they included (a) a limited number of years with sufficient stormflow water-quality data for model calibration and validation, and (b) a lack of selected nutrient and other data for the watershed tributaries and the reservoirs. There also was a lack of data, or readily available data, on tributary or reservoir water-quality and biotic conditions, including (a) bioavailable, rather than just total, phosphorus, (b) total and (or) organic, in addition to inorganic, forms of nitrogen, (c) dissolved, as well as total, organic carbon, and (e) data on chemical, or biological and chemical, oxygen demand.

Monitoring to Address Individual Water-Quality Concerns

The monitoring program and use of monitoring data were reviewed in relation to their ability to address individual long-term and emerging water-quality concerns. For each water-quality concern, this review included an assessment of the ability of RWMA partners or their contractors to describe (a) the state, changes in state or trends, in key water-quality conditions in either the reservoir watershed tributaries or their downstream reservoirs, (b) relate general patterns in land-use conditions or human activities to the water-quality conditions in the reservoir tributaries, and (or) (c) relate tributary water-quality conditions to reservoir water-quality or biotic conditions in the reservoirs.

Long-Term Water-Quality Concerns

Eutrophication accompanied by major algal blooms is a long-term water-quality concern. Long-term monitoring data (approximately 1982–2003) on algal taxa identification and abundance (counts) have been used by RWMA partners to describe seasonal patterns in phytoplankton diversity and production characteristics in each reservoir in relation to eutrophic and mesotrophic conditions. Seasonal patterns of phytoplankton (taxa and abundance), and their die-off and decomposition, have been linked to seasonal patterns in concentrations of DO, hypoxic to anoxic conditions, and concentrations of iron and manganese in reservoir waters. Analysis of time-series data have been used to illustrate apparent declines in algal counts and possibly concentrations of chl-a from the early 1980s through 1990s. Except for one modeling study (described above), there has been limited success relating intra-seasonal variations or annual trends in water-quality conditions in the reservoir watershed tributaries to seasonal or annual trends in water-quality conditions, or phytoplankton production and succession in the reservoirs. Comparisons of phytoplankton data among reservoirs also have been complicated by the lack of data for the upper part of Prettyboy Reservoir, and for major blooms that likely occur in at least the mid-to-lower part of this reservoir in the late winter and early spring.

Eutrophic conditions in the reservoirs, and adjoining portions of selected tributaries supplying the reservoirs, generally are considered to result from excessive nutrients—nitrogen, and in particular, phosphorus—which are long-term water-quality concerns. Extensive monitoring for nitrogen and phosphorus has been conducted in the tributaries, in relation to point and nonpoint sources, and in the reservoirs. Monitoring data from the 1980–90s for nitrogen, exclusively in available forms (ammonia, nitrate, and nitrite), have been used to describe and assess total and form-specific concentrations or loads of nitrogen in the tributaries, their relation to point and nonpoint sources in the reservoir watersheds, and nitrogen concentrations in the reservoirs. Monitoring has documented that wastewater treatment plant (WWTP) upgrades led to the reduction of all forms of available nitrogen in effluents by the late 1990s. Concentrations of total available nitrogen in dry-weather (generally unquantified low) flows also were shown to increase with the extent of agricultural land use, and decrease with the extent of forest cover in a reservoir watershed sub-basin. Seasonal but different patterns were shown to occur in concentrations of nitrate nitrogen in the watershed tributaries and reservoirs. But mid-summer annual lows in tributary dry-weather flows were not observed in the reservoirs, where annual nitrate-nitrogen concentrations actually tend to peak in the spring or early summer. Trend analysis indicated ammonia nitrogen appeared to decline, and nitrate nitrogen appeared to increase, and then possibly leveled off, in dry-weather flows at most tributary stations, whereas only ammonia appeared to decline in the reservoirs. Statistical analysis, limited to Loch Raven and Prettyboy Reservoirs and watersheds, confirmed the trends for ammonia and nitrate nitrogen in the tributaries, but showed that only ammonia nitrogen declined, and at just two stations in Loch Raven Reservoir. The reasons for the differences in nitrate seasonal patterns, or the above trends or lack thereof, in ammonia- and nitrate-nitrogen in the tributaries and reservoirs are unknown. But reservoirs receive both organic (not measured) and inorganic nitrogen loads from storm- (limited sampling after 1995) as well as dry-weather flows. Concentrations of both forms of available nitrogen also generally are lower in the reservoirs compared to dry-weather flows in the tributaries. They also appear to be governed by reservoir stratification, with nitrate nitrogen primarily occurring in the epilimnion and ammonia nitrogen primarily occurring in the hypolimnion. Thus, seasonal patterns, and possibly trends, in available nitrogen likely are determined by nutrient-cycling (abiotic and biotic) and hydrodynamic processes (stratification and turnover) in the reservoirs and not just tributary inflows.

Phosphorus is considered to be the limiting nutrient in phytoplankton production. Monitoring data mainly from the 1980s–90s, primarily for concentrations of total phosphorus (TP), have been used to describe and assess concentrations of phosphorus in the tributaries, their relation to point and nonpoint sources in the reservoir watershed subbasins, and TP concentrations in the reservoirs. Concentrations and total annual loads of TP from three major subbasins in each supply reservoir from the 1980s to the mid-1990s were shown to be highly variable within a year and from year to year presumably because of variations in climate and flow regime. Loads were markedly elevated in wet years compared to dry years. Most of the TP load was carried by stormflow; dry-weather flow accounted for only 15–30 percent of the total annual load depending upon the tributary. Only a relatively small fraction of stormflow TP appeared to be dissolved (orthophosphate) phosphorus (4–21 percent); a higher fraction of the TP in dry-weather flows appeared to be dissolved orthophosphate phosphorus or DOP (75–81 percent). However, concentrations of DOP in stormflows are at least 5 times greater than concentrations of DOP in dry-weather flows.

By the late 1990s, plant upgrades had reduced TP in wastewater effluents by 71–95 percent. For Loch Raven Reservoir, loads from the Western Run subbasin were markedly higher than loads from the other two major subbasins, which indicated that activities to reduce TP loads could focus on this subbasin.

Stormflow and DOP sampling were reduced and discontinued, respectively, by the mid-1990s. Thus, long-term characterizations of the temporal variations of phosphorus in the tributaries, and comparisons of phosphorus in the tributaries and reservoirs were made with TP data primarily obtained from dry-weather flows. During the 1980s–90s, seasonal, but different, patterns in concentrations of TP were observed in the tributaries and reservoirs. Apparent mid-summer peaks in TP in most tributary dry-weather flows were not apparent in the reservoirs, where peak TP concentrations generally appeared in late summer or early fall. Attempts to characterize and compare long-term trends in the concentrations of TP also have proven difficult. Concentrations of TP appeared to decline in dry-weather flows during the 1980s–90s at most tributary stations, but trends could not be statistically verified. Subsequent studies on TP data collected from 1981–2003 revealed that a step change in June or July 1995, and extreme outliers, particularly low TP values in the early data record, complicated trend analysis. The step change could reflect a change in laboratory methods; the outliers could reflect the manner in which immeasurable concentrations of TP data were stored. The documentation of methods and remarking of data, however, are insufficient to confirm either of the above. These types of problems have led to questions on the overall quality of the data in the database (see below). From a broader perspective, the sole reliance on TP analyses coupled with the monitoring of primarily dry-weather flows after 1995, and at least through 2007, likely has and will continue to hinder the ability to estimate loads, to discern patterns and relations among phosphorus concentrations among the tributaries and reservoirs, to describe phosphorus cycling in the reservoirs, and to relate phosphorus to other markers of eutrophication in the reservoirs.

In relation to eutrophication, the frequency of eutrophic compared to mesotrophic conditions in each reservoir is a major water-quality concern, and has been assessed with Carlson Trophic State Indices (TSIs) derived from concentrations of chl-*a*, TP, or DO, and (Secchi disc) transparency data collected from 1982–2000. The frequency of eutrophic compared to mesotrophic conditions observed in the shallow water layer (defined to be 30 feet or less in depth) of each reservoir differed within a given reservoir, and among the reservoirs, depending upon the TSI index used. Among the four TSIs, the TSI chl-*a* index could be a highly sensitive (trend and relative) indicator of the frequency of eutrophic conditions with major algal blooms among the reservoirs. The TSI-DO index also could be a sensitive (trend and relative) indicator of the frequency of hypoxic conditions that result from major algal blooms among the reservoirs. The TSI-transparency index appears to be the least sensitive indicator of eutrophic conditions as the data from 1992–2004 indicate that light penetration is highest in the summer (approaching approximately 20 feet), and lowest in the winter to early spring as a result of lake turnover.

Whereas the analysis of reservoir trophic conditions using the TSI indices indicate that they could help assess the frequency of eutrophic and mesotrophic conditions in the reservoirs, no study was encountered during this review that examined the (1982–2000) trend in results for each index. In addition, there are a number of inadequacies in the monitoring, aggregation, and analyses of the TSI indices data. For most TSI indices, data are collected only monthly throughout the year in Prettyboy Reservoir but bimonthly in the supply reservoirs during periods of reservoir stratification. There also has been no monitoring for chl-*a* data for the upper part of Prettyboy Reservoir, and insufficient monitoring of chl-*a* data for the late winter and early spring in the middle and lower parts of this reservoir. Thus, the TSI index comparisons of the frequency of eutrophic or mesotrophic conditions in this reservoir to the frequency of these conditions in the two supply reservoirs is questionable. Measurements of DO, which serve as the basis of the TSI-DO index comparisons, primarily occur only in the daytime in all three reservoirs. During major algal blooms, these measurements could be positively biased by phytoplankton photosynthesis. Questions concerning the quality of TP data through time already have been noted, which could affect results obtained for the TSI-TP indices—most notably in Loch Raven and Prettyboy Reservoirs. As for transparency, only one-half of the variability in transparency in Loch Raven and Prettyboy Reservoirs can be explained by parameters directly related to light penetration (color, turbidity, and chl-*a*), which is low compared to other studies. In addition, the TSI results for chl-*a*, TP, and DO for the shallow-water layer (defined as 30 feet or less in depth) likely include samples at depths (20 to 30 feet) beyond the determined epilimnion and euphotic layers in at least two of the three reservoirs. Subsequent to the analyses of the TSI-based frequency of eutrophic conditions, the epilimnetic layers, at least for Loch Raven and Prettyboy Reservoirs, were shown to be 15–20 feet in depth for most years of record. In addition, for each of the four TSI indices, the aggregation of sample data over an entire reservoir could mask important spatial differences in the frequency of eutrophic conditions within different parts of each reservoir.

Sedimentation is a long-term water-quality concern. Long-term monitoring in the watershed tributaries has been used to provide data and information on suspended-sediment concentrations, loads, trends, and potential source areas of sediment (among major subbasins). Mean total annual suspended-sediment concentrations and loads markedly varied but were notably higher in wet compared to dry years for each of the three major subbasins monitored in each supply-reservoir watershed during the 1980s through the mid-1990s. More sediment also was carried by stormflow as opposed to dry-weather flow, with the latter accounting for only 10 percent or less of the total annual sediment load. On the basis of areal-weighted dry-weather-flow loads (yields), Beaver Dam and Western Run subbasins, and the lower Gunpowder River subbasin between Prettyboy and Loch Raven Reservoirs were identified as areas for management activities to reduce yields. Suspended-sediment concentrations, and notably loads, also appeared to decline during the mid-1980s. A comparison of suspended-sediment concentration-discharge curves constructed using data from 1982–86 and from 1986–90 for each of three major subbasins in each supply reservoir watershed also indicated less sediment was being transported in each subbasin throughout most of the range in flows after 1985. No statistical analysis was used to confirm the above, nor could these changes be clearly linked to management activities designed to reduce sediment loads as they could reflect a possible change in storm-sampling methods. Since 1995, the reduction in storm sampling has made and will make it difficult to obtain reasonably accurate estimates of annual sediment loads to the reservoirs. Use of dry-weather-flow suspended-sediment data, however, is inadequate as these flows account for very little of the total annual sediment load. After the early 1990s, the magnitude of these flows has been largely unquantified. The transport of sediment to the reservoirs has reduced reservoir storage capacity since their creation. Recent bathymetric surveys (1998–2000) have shown the loss in storage capacity since reservoir construction is 3–11 percent, depending on the reservoir. For Loch Raven Reservoir, which lost 11 percent of its capacity, sediment has filled available storage in the lower parts of the reservoir watershed tributaries and downstream parts of the reservoir. These recent surveys are the starting point for trend assessments on losses in reservoir capacity. Historical (pre-1980s) survey data could not be used because of inadequate documentation of methods and archival of survey data.

Elevated concentrations of manganese and iron in intake waters of both supply reservoirs are a long-term water-quality concern. Monitoring of manganese and iron in intake waters is routinely done by water purveyors only at the water treatment facilities associated with each supply reservoir. Elevated concentrations of these metals have been shown to generally occur following the die-off and decomposition of major algal blooms that contribute to anoxic conditions at depth and the mobilization of these metals from reservoir bed sediments. Monitoring for these metals in the supply reservoirs (bimonthly) is not frequent enough to provide a timely warning of their ultimate occurrence at the intakes in the water-supply reservoirs.

Elevated concentrations of mercury in game fish are a long-term-water-quality concern. Given the initial sources of mercury to the reservoir watersheds and reservoirs appears regional in scope, the concentrations of mercury in game fish occurrence have been largely considered beyond the control of the City or the other RWMA partners. Following droughts that lead to extended withdrawals of water for supply, however, exposed reservoir bed sediments can be covered by dense stands of terrestrial vegetation. Upon reservoir recovery, the submergence of this vegetation could lead to conditions that enhance methylmercury production and biological uptake. Whether or not this occurs cannot be indirectly determined from recent regional knowledge on the occurrence of mercury in game fish in Maryland lakes and reservoirs. Routine synoptic monitoring for mercury in reservoir game fish also cannot address this question. It lacks focus on these reservoir areas prone to bed-sediment exposure and the establishment and eventual submergence of terrestrial vegetation. It also occurs too infrequently to provide suitable comparative data before and after drought events in these areas.

Fecal and other bacteria have been found at elevated counts in selected reservoir watershed tributaries, and are a long-term water-quality concern. The reservoirs, however, appear to function in an effective manner to reduce microbial contaminants at supply intakes. Individual intake sample values and monthly averages for fecal coliform bacteria counts in raw water from Liberty and Loch Raven Reservoirs consistently fall below and well below, respectively, the State water-quality recreational water-contact standard (200 Most Probable Number per milliliter). As of 2007, selected watershed tributaries in each reservoir watershed had bacterial concentrations that exceeded this standard, and the Maryland Department of the Environment requested and is reviewing additional bacterial data from the reservoir watersheds. Long-term monitoring in the reservoir watershed tributaries eventually could require the collection of fecal-coliform or other bacterial data.

Emerging Water-Quality Concerns

Trihalomethanes (THMs) and haloacetic acids (HAAs) are disinfection by-products (DBPs) created by the chlorination of raw supply water, and both are considered to be emerging water-quality concerns. Analysis of monitoring data collected after the late 1990s indicates that DBPs generally form and thus occur in the City water-distribution systems after water treatment. Under a pending rule change, the 30-day moving average concentrations of these DBPs cannot exceed the Federal standards at any monitoring station in either supply reservoir treated water distribution system. An analysis of data from 2003–08 indicates that the total concentration of THMs at one or more distribution monitoring stations exceeded the Federal standard (80 μg/L or micrograms per liter) on approximately 19 percent of the sample-collection dates in both the Loch Raven Reservoir and Liberty Reservoir water-supply distribution systems, and that the total concentration of HAAs exceeded the Federal standard (60 μg/L) at one or more stations on approximately 43 percent and 38 percent of the sample-collection dates in the Loch Raven Reservoir and Liberty Reservoir water-supply distribution systems, respectively. Concentrations of THMs in finished waters also varied seasonally. The highest THM concentrations often occurred from July to September, when algal blooms and die-offs also occur most often in the reservoirs. High THM concentrations also occurred during two drought-recovery periods, which likely also were marked by algal blooms and possibly submerged terrestrial vegetation. There was no correlation between concentrations of THMs and HAAs at a given monitoring station, nor any apparent seasonal trend in HAA concentrations in either water distribution system. Elevated concentrations of HAAs tended to occur at the same monitoring stations in each water distribution system, but at different times in different years. Their occurrence possibly reflects short-term but currently unknown hydrodynamic and (or) biological events. Concentrations of THMs (or HAAs) also bore little relation to concentrations of total organic carbon in raw intake waters, or concentrations of organic carbon after removal of one third of the total carbon in raw intake waters. Hence, monitoring of total carbon alone likely will not identify intake waters that contain elevated concentrations of carbon that form either type of DBP.

Recent increases in the concentrations of sodium and chloride in treated water are emerging water-quality concerns. Monitoring data indicate that sodium concentrations in 2003 were three-to-four-times greater than sodium concentrations in raw waters at the intakes of the supply reservoirs. During the winters of 1999 and 2003, sodium in treated waters from Loch Raven Reservoir often exceeded 20 mg/L (milligrams per liter)—the U.S. Environmental Protection Agency non-regulatory guideline limit for consumers on restricted low-salt-intake diets (500 milligrams per day) diets. During the same period, sodium in treated waters from Liberty Reservoir ranged from 10–16 mg/L, exceeding almost all previous measurements. Chloride concentrations, which have been monitored at supply reservoir intakes since the early 1930s and at tributary dry-weather flow stations since 1982, also have steadily risen in both intake waters and the tributaries throughout these periods. Recently, however, the highest chloride concentrations have occurred during the winter months. The suspected cause for the recent increase in both sodium and chloride concentrations is road-salt use as a deicing agent. As of 2007, however, watershed and reservoir monitoring is too infrequent and lacks inclusion of sodium, and therefore cannot provide a timely assessment of the occurrence of elevated sodium or chloride concentrations in the tributaries or reservoirs before that water reaches the intakes. For the same reason, monitoring likely cannot provide a timely indication of any reductions in sodium or chloride that could result from management activities that could eventually be taken to reduce the use of sodium-chloride salt as a road deicing agent.

Climate variability and climate change could exacerbate withinin-year and year-to-year variations in water-quality conditions associated with most water-quality concerns, and in relation to future monitoring, were addressed as emerging water-quality concerns. Seasonal patterns and year-to-year variations in water-quality conditions associated with long-term concerns, such as eutrophication and sedimentation, as well as emerging concerns, such as DBP and sodium-chloride concentrations, can be linked to the inherent seasonal patterns and annual variability in climate in the Mid-Atlantic region. Recent climate change studies predict increases in both the intensity and frequency of major storms and the duration of periods of hot dry weather for the Mid-Atlantic region. The projected climate changes could imply more frequent drought conditions alleviated by more frequent intense storms with heavy precipitation. Historically, droughts have led to major reservoir withdrawals for supplies, and generally have been followed by a decline in the quality of reservoir water during recovery. The droughts that occurred during the last decade are no exception. The projected climate-related changes described above could result in marked increases in storm-borne contaminants (nutrient, sediment, salt, and bacterial loads) to the tributaries and reservoirs during storm events that lead to reservoir recovery. As of the mid-1990s, the monitoring design, which relies chiefly on dry-weather-flow monitoring, cannot adequately quantify the effects of drought-recovery episodes on tributary and reservoir biotic and water-quality conditions, and thus address most water-quality concerns, particularly if recovery periods increasingly are the result of intense rainfall or snow storms.

Framework Integration to Enhance Water-Quality Monitoring

Collectively, and as of 2007, the reservoir monitoring system could be improved in several key areas to obtain the data necessary to address long-term and emerging water-quality concerns. Improvements could be made in three major areas: (a) the monitoring design framework, (b) the spatial and temporal resolution of water-quality assessments in the major tributaries and reservoirs, and (c) the management and archival of data, including data-quality and methods verification and documentation.

Framework Design

The framework design could include a formal phytoplankton model, such as the Phytoplankton Ecology Group model. This model not only would describe seasonal taxa and their abundance, but intra-seasonal variations in phytoplankton succession. Data from this model could be analyzed in relation to intra-seasonal data on bioavailable (dissolved and total dissolved) as well as total nutrients and trophic indices to (a) improve the RTG's understanding of reservoir water-quality and biotic conditions, (b) help identify nutrient concentrations associated with major algal blooms and their taxa, (c) help identify whether phosphorus and (or) nitrogen are limiting nutrients, and, with time, (d) help identify whether the frequency of eutrophic conditions and (or) occurrence of major algal blooms is declining, and why.

The framework design generally could benefit from alterations in monitoring to improve the characterization of water-quality conditions associated with intra-seasonal and annual variations in climate. This would require monitoring to collect water-quality and biotic data for the full range in flows in tributaries within a year and from year to year.

The framework design could benefit from the aggregation, and as necessary, collection, of the basic hydrodynamic data necessary to quantitatively describe the full range in temporal variations in seasonal and annual climatic and hydrologic conditions for each reservoir watershed and reservoir. The minimal monitoring data required would include daily temperature (mean), daily precipitation (total and type), continuous or partial records of streamflows depending on the type of tributary monitoring station, and daily water levels, withdrawals, and releases from each reservoir.

The framework design generally could benefit from an increased use of statistical methods to help define, aggregate, analyze, and interpret data. Numerous examples where interpretive data appeared in reports but statistical analyses were lacking are presented throughout this review. Although water-quality models have been developed for the Loch Raven and Prettyboy watersheds and reservoirs, these models could be improved, and a similar improved model could be developed for the Liberty watershed and reservoir.

Spatial and Temporal Resolution of Water-Quality Assessments

To illustrate that tributary monitoring falls short in the representation of annual (water year) streamflow in perhaps all but the driest of years, the annual flow regimes for the six gaged tributaries from 2000–06 were examined in relation to daily mean flows, and in particular, the number of days of high flows (75th percentile flows or greater) per year. The period of record chosen reflects several extremely wet as well as dry years. The resultant data indicate that sampling just five storm-flow events per year along with monthly dry-weather flows is likely to cover the annual flow regime in only the driest years of record.

To adequately address tributary flow regime in all three reservoir watersheds, it is proposed that sampling for pre-defined high (or storm-) flow conditions could occur at seven stations (historical stormflow stations in the supply reservoirs and one new station on a tributary to Prettyboy Reservoir) for 3–15 high-flow conditions per station distributed throughout the year. Predefined base-flow conditions would be sampled at each station within a monthly fixed time interval. Sampling at all of the other tributary (remaining historical dry-weather-flow as well as wastewater treatment plant or WWTP) stations would be sampled monthly during a fixed time interval, but regardless of the magnitude of flow condition at each station.

Other changes could be made to improve the spatial and temporal resolution of assessments of water-quality and biotic conditions in the reservoir watershed tributaries and in the reservoirs and thus in relation to modeling or statistical assessments related to most water-quality concerns. Tributary monitoring (including WWTPs) that could address all water-quality and biotic parameter limitations described in this review include in-stream measurements of temperature, pH, DO, conductance, and turbidity. These measurements initially could help determine what methods could be used to collect a representative sample, and ultimately help characterize stream-water-quality conditions at the time of sampling. It also would include the collection at the appropriate times of samples for suspended sediment, sodium, chloride, total and dissolved organic carbon, total-, total dissolved-, and dissolved phosphorus, and total Kjeldahl- or organic-, as well as nitrate plus nitrite-, and ammonia-nitrogen. Reservoir monitoring (generally monthly or bimonthly) at eight historical reservoir-monitoring locations (and as appropriate for the Susquehanna River if it is used) that could address all of the water-quality and biotic parameter limitations noted earlier in this review would include the collection at the appropriate times (gener-ally monthly or bimonthly, but in selected cases, less fre-quently) of samples for suspended sediment, sodium, chloride, total and dissolved organic carbon, total-, total dissolved-, and dissolved phosphorus, and total Kjeldahl- or organic-, as well as nitrate plus nitrite-, and ammonia-nitrogen. In addi-tion, bimonthly sampling would include samples for algal taxa identification and counts and chemical oxygen demand at all eight reservoir-monitoring locations. Sampling at each reser-voir station also would be conducted at depths similar to those used under the ongoing monitoring design, but also at the 15-foot depth during reservoir stratification to ensure sufficient data are obtained in the epilimnion (defined by temperature and transparency conditions).

In addition, a major change that could be implemented with respect to monitoring reservoir water quality could be the use of water-quality profile monitors to provide the short-term data for key parameters necessary to relate watershed tribu-tary-reservoir hydrodynamics and the associated water-quality and biotic conditions that reflect major storm events, and seasonal and annual variations in climate, including drought and recovery conditions. Two fixed-station continuous moni-tors could be established in each reservoir. Each fixed-station

monitor could provide daily 5-foot depth-increment profiles for at least seven parameters—water temperature, DO, pH, specific conductance, chl-*a*, turbidity, and depth of measure-ment. If upgraded to real-time, these reservoir monitors also could provide the data needed to provide advance warning of potential problems in treated waters.

In relation to sedimentation and specifically future bathymetric surveys, the latter could be conducted in the late fall or early spring to reduce interference problems in past surveys caused by submerged aquatic vegetation. The most recent (1998–2000) and future survey results could be checked periodically to verify they are properly documented and that survey data remains properly archived for possible reuse. The most recent survey is the starting point to assess trends in each reservoir. As of 2007, surveys are to be conducted every 10–15 years. Thus, the first real opportunity to determine actual trends in the loss of reservoir storage capacities will be after at least three surveys are completed, which would be sometime in 2030–45.

Data Documentation and Quality

To ultimately reduce the likelihood of future data-quality issues, a comprehensive quality-assurance program and plan (QAPP) plan could be created by the RTG with clear lines of responsibility defined for each QAPP activity. On the basis of the general activities addressed in most standard QAPPs for long-term monitoring programs, such a plan could address all of the limitations identified in the review of the monitoring database and related documentation, through the inclusion of the following:

a) Clear and concise definitions of data-quality require-ments for each hydrodynamic or water-quality param-eter to be obtained;

b) A comprehensive description of the field and labo-ratory methods and procedures used to obtain and provide data, including the required quality-assurance and control data;

c) A comprehensive description of procedures for the archival and remarking of all data;

d) A comprehensive description of procedures for routine annual or bi-annual evaluation of the data collected, including quality-assurance and control data, to deter-mine if data requirements are being met;

e) A comprehensive description of procedures for modi-fying any of the above items (a through d) and docu-menting the changes; and

f) Designation of the organization(s) that will carry out each of the above items.

Implementation of Enhanced Monitoring

The described modifications to the long-term monitoring network for the Baltimore Reservoir System are intended to be comprehensive in nature and address all long-term and emerging water-quality concerns. Prioritization and implementation of selected modifications to the monitoring program would require prioritization of RTG objectives, and ultimately RWMA water-quality concerns. The proposed modifications could be implemented as a pilot or trial effort in one of the supply-reservoir watershed systems. Routine monitoring then could conducted for at least several years, to obtain data over at least a range of climatic conditions, and the resultant data reviewed and analyzed to determine the real benefits compared to the expected benefits. On the basis of this pilot or trial effort, the resultant monitoring design could be implemented throughout the Baltimore Reservoir System.

Acknowledgments

This report is a product of a cooperative investigation between Baltimore City, Carroll and Baltimore Counties, and the U.S. Geological Survey (USGS). To aid in the preparation of this report, interviews were conducted with the following Reservoir Technical Group (RTG) personnel, whom we thank for providing historical and recent technical information related to the reservoir-monitoring program: Ralph Cullison, Chief, Environmental Services Division, Department of Public Works, City of Baltimore Bureau of Water and Wastewater; and within this division, William P. Stack, former Program Administrator, and Robert McAulay, Data Manager, Water-Quality Management Section, Eugene J. Scarpulla, Watershed Manager, Reservoir Natural Resources Section of the Department of Public Works, Baltimore City; Lisa Jones, Analyst, Montebello Water-Quality Laboratory, and Savita Bagel, Analyst, Ashburton Water-Quality Laboratory, Baltimore City Department of Public Works, Baltimore City; Thomas Devilbiss, Deputy Director, Hugh Murphy, Hydrogeologist, Bureau of Resource Management, and James Slater, Compliance Officer, Office of Environmental Compliance, Carroll County Department of Planning, Westminster, Maryland; Stephen Stewart, Manager, Watershed Management Section, Department of Environmental Protection and Sustainability, Baltimore County, Towson, Maryland; Gould Charshee, Chair of the Reservoir Technical Group, Baltimore, Maryland; Timothy C. Rule, Chief, Nutrient and Bacteria Division, Total Maximum Daily Load Technical Development Program, and Charles Poukish, Chief, Biological Assessments Division, Environmental Assessments and Standards Program, Science Services Administration, and Gul Behsudi, Water Resources Engineer, Source Protection and Appropriation Division, Water Supply Program, Water Management Administration, Maryland Department of the Environment, Baltimore, Maryland.

The authors also wish to sincerely thank USGS colleagues W. Reed Green and Douglas B. Chambers, and William P. Stack, Ellicott City, Maryland, for their peer reviews. We also appreciate the reviews of earlier drafts provided by Jerad Bales, USGS, Reston, Virginia, and the following RTG staff or their designees: William P. Stack, Robert McAulay, Thomas Devilbiss, Hugh Murphy, Stephen Stewart, Gould Charshee, and Clark Howells and Duncan Stewart, Department of Public Works, Baltimore County. In addition, the authors thank the following USGS Maryland-Delaware-D.C. Water Science Center (WSC) staff: James M. Gerhart, retired WSC Director, for reviewing and assisting on earlier report drafts, and Robert J. Shedlock and Matthew Pajerowski for their supervisory assistance in arranging meetings with the RTG personnel and staff members of their respective organizations. The authors also thank the following members of the USGS Office of Communications and Publishing, Science Publishing Network: Valerie Gaine, Editor, and Timothy Auer and Gloria Jean Wilson, Visual Information Specialists, who aided in the preparation of this report.

References Cited

Acker, J.G., Harding, L.W., Leptoukh, G., Zhu, T., and Shen, S., 2005, Remotely-sensed chl a at the Chesapeake Bay mouth is correlated with annual freshwater flow to Chesapeake Bay: Geophysical Research Letters, v. 32, L05601, 4 p., DOI:10.1029/2004GL021852.

Amatayakul, P., Cox, R., Defries, R., Delatour, R., Moy, W., Shobrys, D., Boland, J., and Chamberlin, C., 1978, Jones Falls and Loch Raven Watersheds: Baltimore, Maryland, Johns Hopkins University Department of Geography and Environmental Engineering, Final Report to Regional Planning Council, 111 p.

Amatayakul, P., Defries, R., Shobrys, D., Boland, J., and Chamberlin, C., 1978, Jones Falls and Loch Raven Watersheds: Baltimore, Maryland, Johns Hopkins University Department of Geography and Environmental Engineering, Interim Report to Regional Planning Council, 57 p.

Arrigo, K.R., 2005, Marine microorganisms and global nutrient cycles: Nature, v. 437, p. 349–355, DOI:10.1038/nature04159.

Bales, J.D., and Giorgino, M.J., 1998, Lake Hickory, North Carolina: Analysis of ambient conditions and simulation of hydrodynamics, constituent transport, and water-quality characteristics, 1993–94: U.S. Geological Survey Water-Resources Investigations Report 98–4149, 62 p.

Bales, J.D., Sarver, K.M., and Giorgino, M.J., 2001, Mountain Island Lake, North Carolina: Analysis of ambient conditions and simulation of hydrodynamics, constituent transport, and water-quality characteristics, 1996–97: U.S. Geological Survey Water-Resources Investigations Report 01–4138, 85 p.

Baltimore City Department of Public Works, 1992, Appendix C: Reservoir watershed management progress report, Baltimore, Maryland, in Reservoir Watershed Protection Committee, 2000, Action report for reservoir watersheds: Baltimore, Maryland, 70 p., plus appendixes.

Baltimore City Department of Public Works, 1996, Volume I: Selected graphics, report conclusions, and discussion: Baltimore, Maryland, Reservoir Watershed Management Progress Report, 128 p.

Baltimore City Department of Public Works, 2000, Appendix C: Reservoir water quality assessment for Loch Raven, Prettyboy and Liberty reservoirs, Interim Report, Department of Public Works, in Action report 2000 for the reservoir watersheds: Baltimore, Maryland, Watershed Protection Committee, 37 p.

Baltimore City Department of Public Works, 2001, Reservoir water quality assessment for Loch Raven, Prettyboy, and Liberty reservoirs: Baltimore, Maryland, 39 p.

Baltimore County Department of Environmental Protection and Resource Management, 2008, Prettyboy reservoir watershed characterization report, Baltimore County: Towson, Maryland, 101 p.

Baltimore Reservoir Technical Group, 2004, Water quality assessment, targeted studies and ongoing water quality issues in the Baltimore metropolitan water supply reservoirs and their watersheds: Baltimore, Maryland, Reservoir Program Technical Report, 46 p.

Baltimore Reservoir Technical Group, 2007, Implementation of the 2005 reservoir watershed action strategy for calendar year 2006, Partial Progress (Interim) Report: Baltimore, Maryland, Reservoir Technical Group, 11 p.

Banks, W.S.L., and LaMotte, A.E., 1999, Sediment accumulation and water volume in Loch Raven Reservoir, Baltimore County, Maryland: U.S. Geological Survey Water-Resources Investigations Report No. 99–4240, 1 pl., accessed March 15, 2011 at http://md.water.usgs.gov/publications/wrir-99-4240/.

Bergamaschi, B.A., Fram, M.S., Kendall, C., Silva, S.R., Aiken, G.R., and Fujii, R., 1999, Carbon isotopic constraints on the contribution of plant material to the natural precursors of trihalomethanes: Organic Geochemistry, v. 30, no. 8, p. 835–842, DOI: 10.1016/S0146-6830(9a)00066-2.

Bergamaschi, B.A., Kalve, Erica, Guenther, Larry, Mendez, G.O., and Belitz, Kenneth, 2005, An assessment of optical properties of dissolved organic material as quantitative source indicators in the Santa Ana River Basin, Southern California: U.S. Geological Survey Water-Resources Investigations Report 2005–5152, 38 p., available online at http://pubs.usgs.gov/sir/2005/5152/sir_2005-5152.pdf.

Bulgakov, N.G., and Levich, A.P., 1999, The nitrogen: phosphorus ratio as a factor regulating phytoplankton community structure: Archiv für Hydrobiologie, v. 146, no. 1, p. 3–22.

Caffrey, A.J., Hoyer, M.V., and Canfield, D.E., Jr., 2007, Factors affecting the maximum depth of colonization by submersed macrophytes in Florida lakes: Lake and Reservoir Management, v. 23, p. 287–297.

California Water Resources Control Board, 2008, Quality assurance program plan: Sacramento, California, Surface Water Ambient Monitoring Program, Final Technical Report, 190 p.

Carlson, R.E., and Simpson, J., 1996, A coordinator's guide to volunteer lake monitoring methods: North American Lake Management Society, 96 p., accessed March 15, 2011 at http://www.secchidipin.org/tsi.htm.

Chesapeake Bay Program, 2008, Chapter V: Non-tidal water quality monitoring: Annapolis, Maryland, Chesapeake Bay Program, 17 p., accessed March 15, 2011 at http://archive.chesapeakebay.net/pubs/subcommittee/msc/amqawg/Chapter%205%20Nov%2008%20Final.pdf.

Chesapeake Bay Program, 2010, Chesapeake Bay Program split sample program implementation guidelines, Revision 4, 17 p., accessed March 15, 2011 at http://archive.chesapeakebay.net/pubs/quality_assurance/CSSP_Guidelines_12-17-10.pdf.

Downing, J.A., Watson, S.B, and McCauley, Edward, 2001, Predicting Cyanobacteria dominance in lakes: Canadian Journal of Fisheries and Aquatic Sciences, v. 58, no. 10, p. 1,905–1,908, DOI: 10.1139/cjfas-58-10-1905.

Dmytriw, R., Mucci, A., Lucotte, M., and Pichet, P., 1995, The partitioning of mercury in the solid components of dry and flooded forest soils and sediments from a hydroelectric reservoir, Quebec (Canada): Water, Soil, and Air Pollution, v. 80, nos. 1–4, p. 1,099–1,103, DOI: 10.1007/BF01189770.

Fram, M.S., Bergamaschi, B.A., and Fujii, Roger, 2001, Improving water quality in Sweetwater Reservoir, San Diego County, California: Sources and mitigation strategies for trihalomethane (THM)-forming carbon: U.S. Geological Survey Fact Sheet 112–01, 4 p., available online at http://pubs.usgs.gov/fs/fs-112-01/.

Francy, D.S., Myers, D.N., and Helsel, D.R., 2000, Microbiological monitoring for the U.S. Geological Survey National Water-Quality Assessment Program: U.S. Geological Survey Water-Resources Investigations Report 00–4018, 31 p., accessed June 15, 2011 online at *http://oh.water.usgs.gov/reports/wrir/wrir.00-4018.pdf.*

Galloway, J.M., and Green, W.R., 2004, Water-quality assessment of Lakes Maumelle and Winona, Arkansas, 1991 through 2003: U.S. Geological Survey Scientific Investigations Report, 2004–5182, 46 p., available online at *http://pubs.usgs.gov/sir/2004/5182/.*

Galloway, J.M., and Green, W.R., 2006a, Analysis of ambient conditions and simulation of hydrodynamics and water-quality characteristics in Beaver Lake, Arkansas, 2001 through 2003: U.S. Geological Survey Scientific Investigations Report, 2006–5003, 55 p., available online at *http://pubs.usgs.gov/sir/2006/5003/.*

Galloway, J.M. and Green, W.R., 2006b, Application of a two-dimensional reservoir water-quality model of Beaver Lake, Arkansas, for the evaluation of simulated changes in input water quality, 2001–2003: U.S. Geological Survey Scientific Investigations Report 2006–5302, 31 p., available online at *http://pubs.usgs.gov/sir/2006/5302/.*

Gellis, A.C., Banks, W.S.L., Langland, M.J., and Martucci, S.K., 2005, Summary of suspended-sediment data for streams draining the Chesapeake Bay watershed, water years 1952–2002: U.S. Geological Survey Scientific Investigations Report, 2004–5056, 59 p., available online at *http://pubs.usgs.gov/sir/2004/5056/.*

Giorgino, M.J., and Bales, J.D., 1997, Rhodhiss Lake, North Carolina: Analysis of ambient conditions and simulation of hydrodynamics, constituent transport, and water-quality characteristics, 1993–94: U.S. Geological Survey Water-Resources Investigations Report 97–4131, 62 p.

Glodt, S.R., and Pirkey, K.D., 1998, Participation in performance-evaluation studies by U.S. Geological Survey National Water-Quality Laboratory, Denver, Colorado: U.S Geological Survey Fact Sheet FS–23–98, 6 p., accessed January 13, 2010 at *http://nwql.usgs.gov/rpt.shtml?FS-023-98.*

Graham, N.J.D., Wardlaw, V.E., Perry, R., and Jiang, Jia-qian, 1998, The significance of algae as trihalomethane precursors: Water Science and Technology, v. 37, no. 2, p. 83–89.

Granato, G.E., and Smith, K.P., 1999, Estimating concentrations of road-salt constituents in highway-runoff from measurements of specific conductance: U.S. Geological Survey Water-Resources Investigations Report 99–4077, 22 p., available online at *http://pubs.usgs.gov/wri/wri99-4077/.*

Helsel, D.R, and Hirsch, R.M., 2002, Statistical methods in water resources: U.S. Geological Survey Techniques of Water-Resources Investigations, book 4, chap. A3, 522 p.

Hem, J.D., 1970, Study and interpretation of the chemical characteristics of natural waters, 2d ed.: U.S. Geological Water Supply Paper 1473, Washington, D.C., U.S. Government Printing Office, 383 p.

Hirsch, R.M., Moyer, D.L., and Archfield, S.A., 2010, Weighted regressions on time, discharge, and season (WRTDS), with an application to Chesapeake Bay river inputs: Journal of the American Water Resources Association, v. 46, no. 5, p. 857–880.

Intergovernmental Panel on Climate Change, 2007, Climate Change 2007: Synthesis report summary for policymakers: Intergovernmental Panel on Climate Change Plenary XXVII, Valencia, Spain, November 12–17, 2007, 22 p.

Intergovernmental Task Force on Monitoring Water Quality, 1995, The strategy for improving water-quality monitoring in the United States—Final Report of the Intergovernmental Task Force on Monitoring Water Quality: U.S. Geological Survey Open-File Report 95–742, 25 p.

Interstate Commission on the Potomac River Basin, 2006, Modeling framework for simulating hydrodynamics and water quality in the Prettyboy and Loch Raven reservoirs, Gunpowder River basin, Maryland: Rockville, Maryland, Interstate Commission on the Potomac River Basin, 230 p., plus appendixes.

Jack, J., Sellers, T., and Bukaveckas, P.A., 2002, Algal production and trihalomethane formation potential: an experimental assessment and inter-river comparison: Canadian Journal of Fisheries and Aquatic Sciences, v. 59, no. 9, p. 1,482–1,491.

Kaushal, S.S., Groffman, P.M., Band, L.E., Shields, C.A., Morgan, R.P., Palmer, M.A., Belt, K.T., Swan, C.M., Findlay, S.E.G., and Fisher, G.T., 2008, Interaction between urbanization and climate variability amplifies watershed nitrate export in Maryland: Environmental Science & Technology, v. 42, no. 16, p. 5,872–5,878, DOI: 10.1021/es800264f.

Kaushal, S.S., Pace, M.L., Groffman, P.M., Band, L.E., Belt, K.T., Mayer, P.M., and Welty, C., 2010, Land use and climate variability amplify contaminant pulses: Eos Transactions, American Geophysical Union, v. 91, no. 25, p. 221–222, DOI: 10.1029/2010EO250001.

KCI Technologies, Inc., 2004, Total phosphorus in three Baltimore City reservoirs, variations and trends: Baltimore, Maryland, 35 p.

Keith, L.D., 2006, Environmental monitoring and measurement advisor, Instant Reference Sources, Inc., Monroe, Georgia, accessed March 15, 2010 at *http://www.emma-expertsystem.com/.*

Krabbenhoft, D.P., and Fink, L.E., 2001, The effect of dry down and natural fires on mercury methylation in the Florida Everglades, Appendix 7–8 *in* Everglades Consolidation Report: West Palm Beach, Florida, South Florida Watershed Management District, 14 p., accessed March 15, 2010 at *http://my.sfwmd.gov/portal/page/portal/pg_grp_sfwmd_sfer/portlet_prevreport/consolidated_01/chapter%2007/chapter%207%20appendices/a07-08.pdf.*

Krabbenhoft, D.P., Fink, L.E., Olsen, M.L., and Rawlick, P.S., II, 2000, The effect of dry down and natural fires on mercury methylation in the Florida Everglades, *in* Proceedings of the International Conference on Heavy Metals in the Environment, August 6–10, 2000, Ann Arbor, Michigan.

Krasner, S.W., Weinberg, H.S., Richardson, S.D., Pastor, S.J., Chinn, R., Sclimenti, M.J., Onstad, G.D., and Thurston, A.D., Jr., 2006, Occurrence of a new generation of disinfection byproducts: Environmental Science & Technology, v. 40, no. 23, p. 7,175–7,185, DOI: 10.1021/es060353j.

Kraus, T.E.C., Anderson, C.A., Morgenstern, Karl, Downing, B.D., Pellerin, B.A., and Bergamaschi, B.A., 2010, Determining sources of dissolved organic carbon and disinfection byproduct precursors to the McKenzie River, Oregon: Journal of Environmental Quality, v. 39, no. 6, p. 2,100–2,112, DOI: 10.2134/jeq2010.0030.

Kraus, T.E.C., Bergamaschi, B.A., Hernes, P.J., Doctor, D., Kendall, C., Downing, B.D., and Losee, R.F., 2011, How reservoirs alter drinking water quality: Organic matter sources, sinks, and transformations: Lake and Reservoir Management, v. 27, no. 3, p. 205–219.

Kraus, T.E.C., Bergamaschi, B.A., Hernes, P.J., Spencer, R.G.M., Stepanauskas, R., Kendall, C., Losee, R.F., and Fujii, R., 2008, Assessing the contribution of wetlands and subsided islands to dissolved organic matter and disinfection byproduct precursors in the Sacramento-San Joaquin River Delta: A geochemical approach: Organic Geochemistry, v. 39, no. 9, p. 1,302–1,318, DOI: 10.1016/j.orggeochem.2008.05.012.

Langland, M.J., Moyer, D.L., and Blomquist, Joel, 2007, Changes in streamflow, concentrations, and loads in selected nontidal basins in the Chesapeake Bay watershed, 1985–2006: U.S. Geological Survey Open-File Report 2007–1372, 76 p., available online at *http://pubs.usgs.gov/of/2007/1372/.*

Levich, A.P., 1996, The role of nitrogen-phosphorus ratio in selecting for dominance of phytoplankton by cyanobacteria or green algae and its application to reservoir management: Journal of Aquatic Ecosystem Stress and Recovery (formerly Journal of Aquatic Ecosystem Health), v. 5, no. 1, p. 55–61, DOI: 10.1007/BF00691729.

Lindenmayer, D.B., and Likens, G.E., 2009, Adaptive monitoring: a new paradigm for long-term research and monitoring: Trends in Ecology and Evolution, v. 24, no. 9, p. 482–486.

Maryland Department of the Environment, 2002a, Total maximum daily load for mercury in Loch Raven reservoir, Baltimore County, Maryland, Final Report and Final Comments: Baltimore, Maryland, Maryland Department of the Environment, 55 p.

Maryland Department of the Environment, 2002b, Total maximum daily load for mercury in Prettyboy reservoir, Baltimore County, Maryland, Final Report and Final Comments: Baltimore, Maryland, Maryland Department of the Environment, 55 p.

Maryland Department of the Environment, 2002c, Total maximum daily load for mercury in Liberty reservoir, Baltimore County, Maryland, Draft Report for Comments: Baltimore, Maryland, Maryland Department of the Environment, 50 p.

Maryland Department of the Environment, 2003a, Water quality analysis of heavy metals for the Loch Raven reservoir impoundment in Baltimore County, Maryland, Final Report: Baltimore, Maryland, Maryland Department of the Environment, 22 p.

Maryland Department of the Environment, 2003b, Water quality analysis of heavy metals for the Prettyboy reservoir impoundment in Baltimore County, Maryland, Final Report: Baltimore, Maryland, Maryland Department of the Environment, 20 p.

Maryland Department of the Environment, 2003c, Water quality analysis of chromium and lead for the Liberty reservoir impoundment in Baltimore and Carroll Counties, Maryland, Final Report: Baltimore, Maryland, Maryland Department of the Environment, 20 p.

Maryland Department of the Environment, 2003d, Water quality and management plan for the Loch Raven watershed study area in Baltimore County, Maryland: Annapolis, Maryland, Maryland Department of the Environment, 180 p.

Maryland Department of the Environment, 2004a, Source water assessment for Loch Raven watershed: Baltimore, Maryland, Maryland Department of the Environment, 99 p., plus additional appendixes.

Maryland Department of the Environment, 2004b, Approval and rationale of total maximum daily load for mercury in Loch Raven reservoir, Baltimore County, Maryland: Baltimore, Maryland, Maryland Department of the Environment, 14 p.

Maryland Department of the Environment, 2004c, Approval and rationale of total maximum daily load for mercury in Prettyboy reservoir, Baltimore County, Maryland: Baltimore, Maryland, Maryland Department of the Environment, 14 p.

Maryland Department of the Environment, 2006, Total maximum daily loads of phosphorus and sediments for Loch Raven reservoir and total maximum daily loads of phosphorus for Prettyboy reservoir, Baltimore, Carroll and Harford Counties, Maryland, Final Report, 2006: Baltimore, Maryland, Maryland Department of the Environment, 135 p., plus appendixes.

Maryland Department of the Environment, 2008, Integrated report of surface water quality in Maryland: Baltimore, Maryland, Maryland Department of the Environment, 188 p., plus additional tables.

Maryland Department of Planning, 2000a, Maryland Department of Planning land use and land cover data from year 2000, Annapolis, Maryland, unnumbered table, p. 28, in Maryland Department of Natural Resources, 2002, Liberty Reservoir watershed characterization: Annapolis, Maryland, 61 p., plus additional maps.

Maryland Department of Planning, 2000b, Maryland Department of Planning land use and land cover data from Land Cover Map Series 2000, Annapolis, Maryland, fig. 4, p. 5 in Maryland Department of the Environment, 2007, Total maximum daily loads of phosphorus and sediments for Loch Raven reservoir and total maximum daily loads of phosphorus for Prettyboy reservoir, Baltimore, Carroll and Harford Counties, Maryland, Final Report: Baltimore, Maryland, 42 p.

Mash, H., Westerhoff, P.K., Baker, L.A., Nieman, R.A., and Nguyen, My-Lihn, 2004, Dissolved organic matter in Arizona reservoirs: assessment of carbonaceous sources: Organic Geochemistry, v. 35, no. 7, p. 831–843, DOI: 10.1016/j.orggeochem.2004.03.002.

Miller, W.D., Harding, L.W., Jr., and Adolf, J.E., 2006, Hurricane Isabel generated an unusual fall bloom in Chesapeake Bay: Geophysical Research Letters, v. 33, L06612, 4 p., DOI: 1029/2005GL025658.

Mogil, H.M., and Seaman, K.L., 2009, The climate and weather of Delaware, Maryland, and Washington, D.C.: Weatherwise, July–August, p. 16–24.

Morrel, F.M.M., Kraepiel, A.M.L., and Amoyt, Marc, 1998, The chemical cycle and bioaccumulation of mercury: Annual Review of Ecology and Systematics, v. 29, p. 543–566.

Newman, S., and Pietro, K., 2001, Phosphorus storage and release in response to flooding: implications for Everglades stormwater treatment areas: Ecological Engineering, v. 18, no. 1, p. 23–28, DOI: 10.1016/S0925-8574(01)00063-5.

Ortt, R.A., Jr., Kerhin, R.T., Wells, Darlene, and Cornwell, Jeff, 2000, Bathymetric survey and sedimentation analysis of Loch Raven and Prettyboy Reservoirs: Maryland Geological Survey, Coastal and Estuarine Geology File Report Number 99–4, 56 p., plus appendixes, accessed March 15, 2011 at http://www.mgs.md.gov/coastal/pub/FR99-4.pdf .

Padisák, J., 2004, Phytoplankton, chapter 10, in O'Sullivan, P.E., and Reynolds, C.S., eds., 2004, The lakes handbook: limnology and limnetic ecology, volume 1: Malden, Massachusetts, Wiley-Blackwell Publishing Company, p. 251–308.

Palmer, M.A., Liermann, C.A.R., Nilsson, C., Florke, M., Alcamo, J., Lake, P.S., and Bond, N., 2008, Climate change and the world's river basins: Anticipating management options: Frontiers in Ecology and the Environment, v. 6, no. 2, p. 81–89.

Phillips, G.L., 2010, Eutrophication of shallow temperate lakes, chapter 10, in O'Sullivan, P.E., and Reynolds, C.S., eds., 2004, The lakes handbook: limnology and limnetic ecology, volume 2: Malden, Massachusetts, Wiley-Blackwell Publishing Company, p. 261–278.

Rasmussen, P.P., and McAllister, D.H., 2005, Monitoring the water quality of Lake Olathe, Johnson County, Kansas: U.S. Geological Survey Fact Sheet 2005–3093, 2 p., available online at http://pubs.usgs.gov/fs/2005/3093/.

Rast, W., and Lee, G.F., 1978, Summary of analysis of the North American (U.S. portion) OECD Eutrophication Project: Nutrient loading—Lake response relationships and trophic state indices: Corvallis, Oregon, U.S. Environmental Protection Agency Report Number EPA-600/3-78-008, 453 p.

Rawlick, P., 2001, Stormwater treatment area 1 west: Results of startup mercury monitoring, appendixes 7–14, in Everglades Consolidated Report: West Palm Beach, Florida, South Florida Watershed Management District, 12 p., accessed March 15, 2011 at https://my.sfwmd.gov/portal/page/portal/pg_grp_sfwmd_sfer/portlet_prevreport/consolidated_01/index.html.

Redfield, A.C., 1958, The biological control of chemical factors in the environment: American Scientist, v. 46, p. 205–221.

Reinelt, L.E., Horner, R.R., and Mar, B.W., 1988, Nonpoint source pollution monitoring program design: Journal of Water Resources Planning and Management, v. 114, no. 3, p. 335–352.

Reservoir Watershed Management Agreement, 2005, Baltimore management agreement: Baltimore, Maryland, Reservoir Watershed Protection Program, Baltimore, Maryland, 10 p.

Reservoir Watershed Management Agreement Action Strategy, 2005, Action strategy for the reservoir watersheds: Baltimore, Maryland, Baltimore Reservoir Management Program, Baltimore, Maryland, 13 p.

Reservoir Watershed Protection Committee, 2000, Action report for reservoir watersheds: Baltimore, Maryland, Reservoir Watershed Protection Committee, 70 p., plus additional appendixes.

Reservoir Watershed Protection Subcommittee, 1992, Action report for the reservoir watersheds: Baltimore, Maryland, Reservoir Watershed Protection Subcommittee, 17 p., plus additional appendixes.

Rowland, R.C., Westenburg, C.L., Veley, R.J., and Nylund, W.E., 2006, Physical and chemical water-quality data from automatic profiling systems, Boulder Basin, Lake Mead, Arizona and Nevada, water years 2001–04: U.S. Geological Survey Open-File Report 2006–1284, 27 p., available online at *http://pubs.usgs.gov/of/2006/1284/*.

Ryding, O.E., and Rast, W., eds., 1989, The control of eutrophication in lakes and reservoirs: Paris, France, Man and the Biosphere Series, v. 1, United Nations Educational, Scientific, and Cultural Organization (UNESCO), 314 p.

Sarver, K.M., and Steiner, B.C., 1998, Hydrologic and water-quality data from Mountain Island Lake, North Carolina, 1994–97: U.S. Geological Survey Open-File Report 98–549, 165 p.

Schloss, Jeffrey, 2002, Murky waters: Gaining clarity on water transparency measurements *in* The clean water team guidance compendium for watershed monitoring and assessment: Durham, New Hampshire, University of New Hampshire, Cooperative Extension and the Center for Freshwater Biology, State Water Resources Control Board, Division of Water Quality, 12 p.

Singer, P.C., 1999, Humic substances as precursors for potentially harmful disinfection by-products: Water and Science Technology, v. 40, no. 9, p. 25–30.

Singer, P.C, Weinberg, H.S, Krasner, S., Harish, A., and Najm, I., 2002, Relative dominance of HAAs and THMs in treated drinking water: Denver, Colorado, American Water Works Association and Research Foundation, 293 p.

Sommer, U., Gliwicz, Z.M., Lampert, W., and Duncan, A., 1986, The PEG-model of seasonal succession of planktonic events in fresh waters: Archiv für Hydrobiologie, v. 106, no. 4, p. 433–471.

Stewart, S., Gemmill, E., and Pentz, N., 2005, An evaluation of the functions and effectiveness of riparian forest buffers: Alexandria, Virginia, Baltimore County Department of Environmental Protection and Resource Management, Report to the Water Environment Research Foundation, 123 p., plus additional appendixes.

Susquehanna River Basin Commission, 2006, Conowingo pond management plan: Harrisburg, Pennsylvania, Susquehanna River Basin Commission Publication Number 242, 153 p.

Sveinsdottir, A.Y., and Mason, R.P., 2005, Factors controlling mercury and methylmercury concentrations in largemouth bass (*Micropterus salmoides*) and other fish from Maryland reservoirs: Archives of Environmental Contaminant Toxicology, v. 49, no. 4, p. 528–545.

The H. John Heinz III Center for Science, Economics, and the Environment, 2008, The state of the nation's ecosystems 2008: Focus on nitrogen: Washington, D.C., Fact Sheet, 6 p., accessed June 26, 2011 at *http://www.heinzctr.org/ecosystems/2008report/pdf_files/Nitrogen_Fact_Sheet.pdf*.

Tõnno, I., 1999, The impact of nitrogen and phosphorus concentration and N/P ratio on cyanobacterial dominance and N2 fixation in some Estonian Lakes, Tartu: Estonia, Spain, University Press, 46 p.

U.S. Environmental Protection Agency, 2001, Requirements for quality-assurance project plans EPA QA/R-5: Washington, D.C., Office of Environmental Planning, U.S. Environmental Protection Agency Report EPA/240/B-01/003, 24 p., plus appendixes, accessed March 15, 2011 at *http://www.epa.gov/quality1/qs-docs/r2-final.pdf*.

U.S. Environmental Protection Agency, 2003, Drinking water advisory: Consumer acceptability advice and health effects analysis on sodium: U.S. Environmental Protection Agency, Office of Water, Washington, D.C., Report Number EPA 822-R-03-006, 29 p.

U.S. Environmental Protection Agency, 2006, National Primary Drinking Water Regulations: 40 CFR Parts 9, 141, and 142, Stage 2 Disinfectants and Disinfection Byproducts Rule: Final Rule, *in* Federal Register, National Archives and Records Administration, v. 72, no. 2, p. 392–436, accessed March 15, 2011 at *http://frwebgate. access.gpo.gov/cgi-bin/getdoc.cgi?dbname=2006_ register&docid=fr04ja06-14.pdf.*

U.S. Environmental Protection Agency, 2008, National Risk Management Research Laboratory's quality assurance program plan requirements for measurement projects, 2 p., accessed March 15, 2011 at *http://www.epa.gov/nrmrl/qa/ pdf/MeasurementQAPPNRMRLrev0.pdf.*

U.S. Environmental Protection Agency, 2009, Drinking water standards and health advisories, 2009 edition: Washington, D.C., U.S. Environmental Protection Agency, Office of Water, Report Number EPA 822-R-09-011, 12 p.

U.S. Geological Survey, variously dated, National field manual for the collection of water-quality data: U.S. Geological Survey Techniques of Water-Resources Investigations, book 9, chaps. A1–A9, available online at *http://pubs.water.usgs. gov/twri9A.*

Valcik, J.A., 1975, Algae in Baltimore's reservoirs: Journal of American Water Works Association, v. 67, no. 3, p. 109–113.

Vitousek, P.M., Mooney, H.A., Lubchenco, Jane, and Melillo, J.M., 1997, Human domination of Earth's ecosystems: Science, v. 277, no. 5325, p. 494–499.

Wagner, R.J., Bougler, R.W., Jr., Oblinger, C.J., and Smith, B.A., 2006, Guidelines and standard procedures for continuous water-quality monitors: Station operation, record computation, and data reporting: U.S. Geological Survey, Techniques and Methods Report 1–D3, 51 p., plus attachments, available online at *http://pubs.usgs.gov/tm/2006/ tm1D3/.*

Walker, W.W., Jr., 1988, Evaluating watershed monitoring programs, Report to Bureau of Water and Waste Water: Baltimore, Maryland, 87 p., plus additional appendix.

Wall, G.R., Nystrom, E.A., and Litten S., 2008, Suspended sediment transport in the freshwater reach of the Hudson River estuary in eastern New York: Estuaries and Coasts, v. 31, no. 3, p. 542–553, DOI: 10.1007/s12237-008-9050-y.

Weisberg, S.B., Rose, K.A., Clevenger, B.S., and Smith, J.O, 1985, Chapter III: Dams with potential for hydroelectric power, *in* Inventory of Maryland Dams and Assessment of Hydropower Resources: Columbia, Maryland, Martin Marietta Environmental Systems, Report Number PPSP-85-02, p. 4–9.

Weiss, C.M., and Kuenzler, E.J., 1976, The trophic state of North Carolina lakes: Chapel Hill, North Carolina, Water Resources Research Institute of the University of North Carolina, Report Number 119, 224 p.

Wetzel, R.G., 2001, Limnology: lake and river ecosystems, 3d ed.: New York, Academic Press, Elsevier Science Imprint, 1,006 p.

Wilde, F.W., ed., 2006, Collection of water samples (ver. 2.0): U.S. Geological Survey Techniques of Water-Resources Investigations, book 9, chap. A4, September 2006, available online at *http://pubs.water.usgs.gov/twri9A4/.*

Winfield, G.L., and Sakai, J., 2003, Task 6: Final Liberty watershed assessment: Baltimore, Maryland, Department of Public Works, Bureau of Water and Wastewater, Project No. 812 On-Call Engineering Services 110 p., plus additional appendix.

Winfield, G.L., Sakai, J., and Stack, W., 2006, Road salt contamination of the Baltimore metropolitan reservoirs, Potomac River Drinking Water Source Protection Partnership Meeting Proceedings, February 15, 2006, Interstate Commission on the Potomac River Basin, Rockville, Maryland, 18 figs.

Wu, W.W., Chadik, P.A., and Davis, W.M., 2000, The effect of structural characteristics of humic substances on disinfection by-product formation in chlorination, chapter 8, p. 109–121, *in* Barrett, S.E., Krasner, S.W., and Amy, G.L., eds., 2000, Natural organic matter and disinfection byproducts: Characterization and control in drinking water: Washington, D.C., American Chemical Society Symposium Series 761, 163 p.

Xie, Y.F., 2004, Disinfection byproducts in drinking water: Formation, analysis, and control: Boca Raton, Florida, Lewis Publishers, CRC Press, 160 p.

Appendix A: Water-Quality Monitoring to Support Watershed Restoration

Contents

Figure

Table

Introduction

Baltimore City is primarily responsible for managing and monitoring water-quality conditions in the City water-supply reservoirs, and in selected tributaries in the reservoir watersheds. Because the reservoir watersheds lie largely outside the jurisdiction of the City, however, managing and assessing reservoir-watershed conditions that could affect reservoir water quality is shared by City, County, and State governments.

Management of the reservoir watersheds has been guided by characterizations and source-water assessments that reflect the environmental state of the reservoir watersheds and their tributaries. These characterizations and assessments have been helped or were followed by short-term and synoptic studies conducted on tributary streams in each reservoir watershed to identify nonpoint pollutant problems and their source areas, to develop and implement stream restoration strategies, and to improve tributary water and habitat quality and reduce pollutant loads. These activities are described using representative examples below. In each case, the descriptions are illustrative of these activities and are not a comprehensive summary of all activities that have been conducted in the reservoir watersheds by agencies of the State of Maryland or Baltimore and Carroll Counties.

Source-Water Assessments

Amendments to the 1996 Safe Drinking Water Act (SWDA) required states to conduct source-water assessments to evaluate the safety of all public drinking-water systems. These assessments are to include a comprehensive characterization of each reservoir watershed as well as the reservoir to better enable strategies to be developed to maintain or improve the quality of water, biota, and habitat in streams and reservoirs. The source-water assessment studies for both the Liberty and Loch Raven Reservoir watersheds were completed in 2003 and 2004, respectively (Winfield and Sakai, 2003; Maryland Department of the Environment, 2004). As part of these assessments, water-quality data from the core monitoring program were analyzed and summarized to describe point and nonpoint sources of pollutants. Information from these assessments has been used throughout this retrospective review.

Relevant to the retrospective review of the monitoring program, source-water assessments provided time-of-travel studies. These studies indicate that low flows can travel from the headwaters of tributaries to the reservoirs within approximately half a day to 2 days. High flows, however, such as those associated with storms, likely reach a reservoir fairly quickly, in less than a quarter to a half day.

The rapid traveltimes of storms have implications for long-term monitoring. Unless a storm runoff event is large enough to displace a major portion of the stored reservoir water, and in-lake monitoring occurs shortly afterwards, the direct impact of the storm on the reservoir is not measured.

As noted by the Interstate Commission on the Potomac River Basin (2006), in-lake monitoring by the Reservoir Watershed Management Agreement (RWMA) partners is simply too infrequent (monthly to bi-monthly). By design, storm-related reservoir sampling also generally does not occur if reservoir conditions are unsafe for sampling, such as during storms. Thus, the probability of capturing the impact of a major storm-runoff event on the reservoirs is low. When a storm occurs and at least some of its effects on reservoir water quality are measured, the resultant data complicate analyses of long-term trends as well as modeling of water quality in the reservoir because few events are in fact adequately captured.

A consequence of the current reservoir monitoring is that the influence of storms on reservoir water quality, such as sedimentation and turbidity, nutrient enrichment (phosphorus), and changes in dissolved-oxygen concentrations, in both the epilimnetic and hypolimnetic layers also are poorly understood, and therefore, difficult to accurately model (Interstate Commission on the Potomac River Basin, 2006). The ability to obtain water-quality data for storm events in both the watershed tributaries and reservoirs is desirable to accurately assess conditions of state, changes in state, trends, and loads, and improve modeling to determine whether progress is being made towards addressing RWMA partner water-quality concerns and technical goals.

Other aspects of the source-water assessments that involve monitoring to address RWMA partner concerns, and achieve RWMA goals, are short-term studies conducted by RWMA partners that focus on the watershed tributaries. These studies are designed to describe the conditions of streams, to identify impaired streams, to identify actions needed to reduce impairments, and to prioritize impaired streams for restoration. They also are important to the large-scale long-term monitoring program. The source-water assessments include identification of areas impaired by agricultural or urban development. For those areas that are upstream of the long-term monitoring stations, their restroation could lead to detected improvements in tributary water-quality conditions. In addition, source-water assessments can help identify impaired areas, such as eroded streambeds and banks, or degraded forest areas, for restoration, that generally would not be identified and restored by use of traditional best-management practices (BMPs) that focus on agricultural or urban land use and land owners, which could also lead to improvements in monitored reservoir watershed tributary conditions.

Watershed Characterization Studies

As part of the source-water assessment studies, watershed-characterization studies have been conducted on the reservoir watersheds (Maryland Department of Natural Resources, 2002a; Maryland Department of the Environment, 2003; Baltimore County Department of Environmental Protection and Resource Management, 2008). These studies were conducted to partially fulfill Federal requirements under the National Storm Discharge Elimination Site (NPDES) Municipal Stormwater Discharge Permit, and State programs, such as the 1998 Maryland Clean Water Action Plan (Maryland Department of Natural Resources, 1998). Under the latter, the Liberty and Loch Raven-Prettyboy Reservoirs and their watersheds were designated as watersheds within the state that have the highest priority for protection and restoration, and warrant a comprehensive watershed-restoration action plan.

As part of each watershed characterization, information was compiled on land use and land cover and known or potential point and nonpoint sources of pollutants. Most nonpoint sources of pollutants actually are defined in relation to land use and land cover, on the basis of early studies that compared different types of land use and land cover to water-quality data obtained from the long-term tributary dry-weather monitoring program (Baltimore City Department of Public Works, 1996, 2000, 2001). In addition, short-term monitoring and synoptic studies were used to help characterize water quality in relation to land use within selected reservoir watershed subbasins. The selected subbasins include subbasins identified by the long-term monitoring network as source areas for elevated nutrient and sediment loads, as well as subbasins not covered by the long-term monitoring network—for example, within close proximity to the reservoirs. Examples of monitoring data collected, and information provided from the analysis of these data, are provided as part of this retrospective to show (a) that impairment conditions in tributary streams directly relate to RWMA goals and water-quality concerns, and (b) that the location of impaired streams and plans to restore impaired streams have a bearing on the design of the long-term monitoring network.

Short-Term Stream Monitoring Studies

Short-term monitoring (typically 1–2 years) of stream-water quality is conducted by RWMA partners to address water-quality conditions in the small subbasins within each reservoir watershed. Using the Loch Raven Reservoir watershed as an example, short-term (1-year) monitoring was conducted at over two dozen stream sites in the lower part of this watershed in 1998 (fig. A1). The monitoring stations were used in part to assess source-water conditions, and most were established around the periphery of the reservoir in subbasins not covered by the long-term monitoring network in the reservoir watersheds (see **Main Report**, fig. 3). Selected monitoring stations were located in Piney Run primarily to address concerns with effluent discharge by the wastewater treatment plant (WWTP) at Hampstead, Maryland.

The purposes of this monitoring network were to provide data to: (a) address potential risks to drinking-water quality and aquatic health from pollutants—nutrients, metals, and bacteria or other pathogens—in stream base- and stormflows;

(b) calibrate a model to estimate loads by land use, to further aid the identification of excessive pollutant-source areas; and (c) for Piney Run, to determine if the Hampstead WWTP effluent impacted downstream water quality and quantity. From the monitoring data, the State and Baltimore County (Maryland Department of the Environment, 2003) determined that:

a) Subbasins where development reflected urban and residential land use produce elevated base and stormflows relative to rural subbasins with agricultural or forested land use.

b) Subbasins with developed land use had elevated nitrogen and phosphorus concentrations in base and stormflows compared to subbasins with appreciable forest cover. Nitrate concentrations were elevated at 13 of 15 sites under low-flow conditions, indicating widespread contamination of groundwater.

c) Selected agricultural and urban sites had high concentrations of fecal coliform bacteria during low flow, which was attributed to livestock operations at the agricultural sites, but elevated concentrations also were found at selected urban sites for unknown reasons. *Cryptosporidium* was not detected. *Giardia* cysts were detected by presumptive and definitive tests, but at concentrations well below levels related to an infectious dose of 150 cysts.

d) Arsenic, barium, chromium, and nickel did not exceed water-quality standards during base or stormflows, but two metals (copper and lead) were considered a potential threat to aquatic life at two locations, likely as a result of livestock operations, with the highest concentrations occurring during stormflows.

e) Stream pH was within the acceptable range (6.5–8.5 standard units), except in areas underlain by limestone karsts, where base-flow pH was likely to exceed 8.5.

f) Concentrations of dissolved oxygen typically were above the 5.0 mg/L (milligrams per liter) RWMA standard. Biological oxygen demand, chemical oxygen demand, and total organic carbon concentrations indicated little potential for oxygen depletion in these streams.

g) Although storm data were limited, atrazine concentrations did not appear to pose a health risk in either base or stormflows.

h) Hampstead WWTP effluent constituted approximately 82 percent of base flow in the headwaters of Piney Run and effluent quality determined stream-water quality downstream. During recorded stormflows (peak discharges of 20–40 ft^3/s, or cubic feet per second), the WWTP flow was only 4 ft^3/s, and had considerably less influence on stream-water quality.

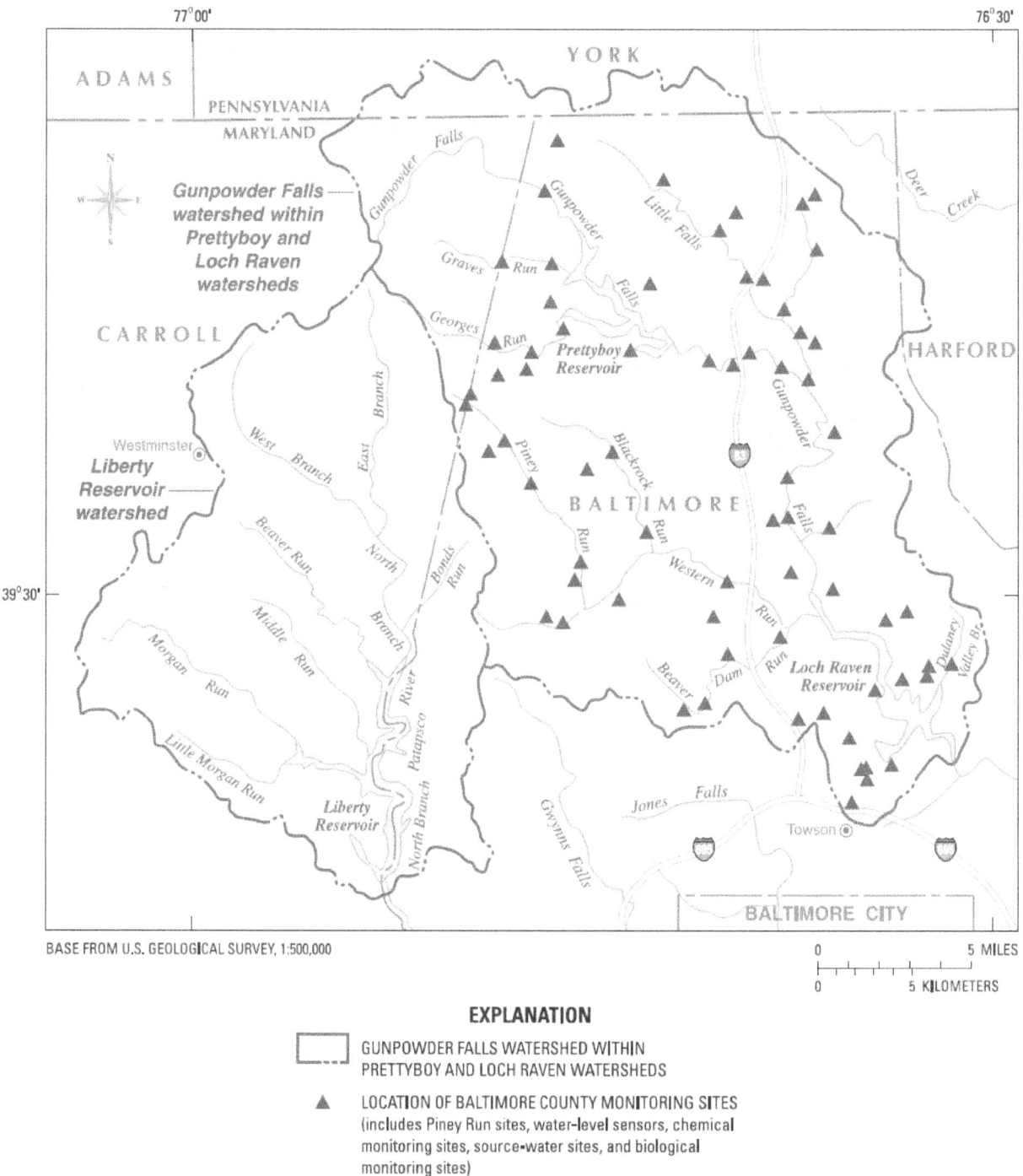

Figure A1. Gunpowder Falls watershed monitoring sites within Loch Raven and Prettyboy Reservoir watersheds (modified from Baltimore County Department of Environmental Protection and Sustainability, formerly Baltimore County Department of Environmental Protection and Resource Management, 1998).

Stream Surveys

Three types of stream surveys have been used to characterize water-quality conditions in subbasins in the Baltimore Reservoir watersheds, and include synoptic surveys for nutrients (nitrogen and phosphorus), stream corridor and stability surveys, and stream habitat and biological surveys.

Synoptic Surveys for Nutrients

Spring low-flow nutrient synoptic surveys, in lieu of short-term monitoring, are another method used to determine relations between land use and land cover (human activities) and nutrient loads, and help identify pollutant source areas on a subbasin scale. By design, these surveys were conducted in the spring—in April 2002 for the Carroll County parts of the Liberty Reservoir watershed (Maryland Department of Natural Resources, 2002b) and in April 2006 for the Prettyboy Reservoir watershed (Maryland Department of the Environment, 2006a). Although total phosphorus concentrations are not highest in reservoir tributaries in the spring, nitrate concentrations and dry-weather flows are on average highest in April. Thus, nitrogen and phosphorus dry-weather loads were expected to be at or near their annual highs in April.

Within each watershed, synoptic data were used to describe and compare low-flow nutrient concentrations and loads among subbasins (table A1). For the Liberty Reservoir watershed, and according to synoptic-survey criteria, among the 41 synoptic sites, all sites exhibited elevated nitrogen—59 percent (25/41) had high to excessive nitrogen concentrations, and 15 percent (6/41) of the sites also had high to excessive (instantaneous) areal-weighted nitrogen loads or yields (Maryland Department of Natural Resources, 2002b). For orthophosphate-phosphorus, only 8 percent of the survey sites had concentrations that were considered high to excessive, but none of the sites had phosphorus loads that exceeded baseline conditions (0.0005 kilograms per hectare per day or less). The Middle Run and Western Run subbasins had the greatest number of internal subbasins with high to excessive low-flow nutrient concentrations. In addition, most sites in the four sub-watersheds covered by this survey (which also included Snowden Run and Roaring Run tributaries) that had high to excessive nutrient concentrations were in developed headwater subbasins.

For the Prettyboy Reservoir watershed, all but 1 of the 68 nutrient synoptic sites exhibited excessive nitrate-nitrite concentrations at low flows compared to the survey baseline standard, and most (88 percent or 60/68) sites had concentrations that were considered high to excessive

Table A1. Nutrient synoptic summaries for Liberty and Prettyboy Reservoir watersheds (from Maryland Department of Natural Resources, 2002b and Maryland Department of the Environment, 2006a).

[%, percent, equals the ratio of the number of sites with either baseline to moderate, or high to excessive, nitrate-nitrite or orthophosphate concentrations (or loads) to total number of synoptic sites, multiplied by 100; N, nitrogen; P, phosphorus; <, less than; >, greater than; mg/L, milligrams per liter; kg/ha/d, kilograms per hectare per day]

Reservoir/ watershed	Date of synoptic	Synoptic site distribution	Number of synoptic sites	Proportion of total number of synoptic sites (%)							
				With nitrate-nitrite N concentrations within the specified range		With nitrate-nitrite N areally weighted loads within the specified range		With orthophosphate P concentrations within the specified range		With orthophosphate P areally weighted loads within the specified range	
				Baseline to moderate (<1 to 3 mg/L)[1]	High to excessive (3.1 to >5 mg/L)[2]	Baseline to moderate (<0.01 to 0.02 kg /ha/d)[1]	High to excessive (0.021 to >0.03 kg / ha/d)[2]	Baseline to moderate (<0.005 to 0.010 mg/L)[1]	High to excessive (0.011 to >0.015 mg/L)[2]	Baseline to moderate (<0.0005 to 0.001 kg/ha/d)[1]	High to excessive (0.0015 to >0.003 kg/ha/d)[2]
Liberty	April 2002	Among four subbasins	41	41	59	85	15	70	8	100	0
Prettyboy	April 2006	Throughout watershed	68	12	88	4	96	93	7	100	0

[1] First value defines baseline concentrations, which are less than the specified numerical value; moderate values lie between the defined upper threshold for baseline values up to the second value specified. Ranges were defined by Frink (1991) for the Chesapeake Bay watershed.

[2] High concentrations are those that occur at the first value and up to the second value; excessive concentrations are those that exceed the second value. Ranges were defined by Frink (1991) for the Chesapeake Bay watershed.

(table A1). Furthermore, approximately 96 percent of the sites had basin-area-weighted-flow (instantaneous) nitrogen loads that were excessive. Only about 8 percent of the subbasins had high to excessive orthophosphate-phosphorus concentrations, and none of the sites had high to excessive phosphorus loads. Subbasins with high to excessive nitrate-nitrite concentrations often were clustered together, and chiefly occurred in two sub-basins—Georges Creek and Prettyboy Branch sub-watersheds (Maryland Department of the Environment, 2006a). High to excessive nutrient concentrations were associated with agricultural and developed subbasins, mainly row-crop and livestock agriculture and low-density residential communities on septic systems.

Collectively, the synoptic surveys were shown to be useful in the identification of subbasins within each reservoir watershed that were potential source areas for high to excessive nutrient (primarily nitrogen) concentrations and, in some cases, nutrient loads, at low flows. Results from the two synoptic surveys, however, cannot be compared to prioritize subbasins among reservoir watersheds as the synoptic in each reservoir watershed occurred in different years with different hydrologic conditions. The Liberty Reservoir watershed synoptic was conducted in 2002 during a very dry spring. Surveyed low-flow nutrient concentrations were lower than the typical annual averages for streams in this and other watershed areas (Maryland Department of Natural Resources, 2002b). The Prettyboy Reservoir watershed synoptic occurred in 2006 during a very wet spring. Surveyed low-flow nutrient concentrations were higher than typical averages for streams in this and other watershed areas (Maryland Department of the Environment, 2006a). For reasons similar to those described above, results cannot be combined for the synoptic and short-term monitoring. The short-term monitoring in the Loch Raven Reservoir watershed was conducted in 1988, a relatively dry year.

Stream Corridor and Stability Surveys

In addition to nutrient surveys, stream corridor and stability surveys have been developed and used to assess the impact of land use and land cover (human activities) on the physical condition of streams in the reservoir watersheds, and aid in the development of RWMA watershed restoration action strategies and priorities (Maryland Department of Natural Resources, 2002c; Maryland Department of the Environment, 2004, 2006b). Collectively, these surveys have provided a wealth of data and information in relation to both of these objectives, including the following:

a) Information on the occurrence, extent, and possible causes of observed instabilities in third- and lower-order streams, including bed incision or aggregation and bank erosion (widening or mass wasting) or deposition, and the potential for continued erosion and, if performed, effective restoration;

b) Information on the integrity of the riparian zone adjacent to the stream, including the type, width, density, and appearance of vegetation; and

c) Information on stream-corridor biotic and water-quality indicators, including physical habitat, and the occurrence and conditions that result from stormwater BMPs, storm-drain outfalls, roadways, construction, exposed sewer lines, non-permitted discharges, and trash or dumping.

Corridor and stability surveys used in the Baltimore reservoir watersheds differed in that the former primarily obtained information through visual observation, were more qualitative than quantitative in nature, and thus enabled coverage of a greater number of streams in a subbasin than the latter. Stream-corridor studies generally were conducted before stream-stability surveys over large areas of the reservoir watersheds. The information obtained from corridor studies allowed the RWMA partners to make general comparisons of stream conditions among major, and within a major, minor, subbasins within the reservoir watersheds. In this regard, they helped identify small subbasins within a major subbasin whose stream corridors appeared to be impaired.

On the basis of information obtained from the stream-corridor surveys, stream-stability surveys were developed mainly to further examine streams in the subbasins. The subbasins surveyed for stability generally had a high frequency of potentially moderately to highly impaired water-quality conditions. The objectives of stream-stability surveys were to provide detailed information on: (a) the current morphological states of the small (generally first- and second-order) stream corridors within a targeted subbasin, (b) the likelihood these streams would maintain their current morphology or undergo a change in morphology, and (c) if their morphological condition was unstable, whether or not stream restoration was warranted, and what it likely would require.

Stability surveys appear to be an effective monitoring tool for the RWMA partners to help determine what restorative actions on which streams would be most effective in a surveyed subbasin. Because of the quantitative data requirements of stream-stability surveys, however, they generally have been conducted in only a few major subbasins in each reservoir watershed, and within each major subbasin, generally on some but not all small subbasins within a major subbasin.

Collectively, the stream-corridor and stability surveys (Maryland Department of Natural Resources, 2002c; Maryland Department of the Environment, 2004, 2006b) have shown that highly unstable (eroding) and chiefly first- and second-order stream corridors occur in a variety of different but mostly headwater settings in surveyed subbasins in all three reservoir watersheds. These settings range from reaches without any riparian (forested) buffer lying adjacent to developed lands to the presence of more than adequate riparian buffers that are subject to inadequately controlled stormwater runoff. Where stream erosion is observed, channels are most often undergoing incision, or if already incised, are widening through bank cutting; many of the stream corridors undergoing degradation (incision, widening, or both) are in upland headwater basins, and these eroding streams likely are a major source of sediment to downstream tributaries and inevitably the watershed reservoirs. The RWMA partners have identified and prioritized stream reaches for restoration activities throughout the surveyed subbasins in each reservoir watershed.

Stream Habitat and Biological Surveys

Two types of monitoring surveys have been used to assess the habitat conditions related to the biotic health of reservoir watershed tributary streams. Initial stream-habitat assessments were frequently conducted as part of the stream-corridor and stability studies. The results from these surveys have provided site-specific information that is useful in the characterization of the suitability of streams to support designated uses related to recreational fishing, and to sustain native and stocked trout populations.

Stream habitat surveys identified fish-migration barriers. Barriers most often consisted of debris blockages or limited flow and depth conditions, but included human-constructed structures that could interfere with fish migration.

Although information was not available for the lower Loch Raven Reservoir watershed, during stream corridor surveys for the Liberty Reservoir watershed, survey crews identified 32 such barriers (Maryland Department of Natural Resources, 2002c). The majority of these barriers (22/32) blocked the entire width of the stream. Artificial barriers (23) dominated, and included dams (9), pipe crossings (5), road crossings (5), concrete debris (3), and a streamgage. Natural barriers (9) included beaver dams (3), natural falls (3), and an in-stream pond, a channelized stream, and a large rock. On a sub-watershed basis, West Branch, Middle Run, and Snowdens Run had 18, 9, and 5 barriers, respectively.

For the Prettyboy Reservoir watershed, stream-corridor survey crews identified 17 fish migration barriers (Maryland Department of Natural Resources, 2006b). Most of the barriers were artificial (12/17), and included road crossings (10) and dams (2). Natural barriers consisted of natural falls (2) and debris dams (2).

Stream-stability surveys also provided a more detailed assessment of stream characteristics related to biological habitat for generally small (first- and second-order) streams. On the basis of surveys conducted in the lower Loch Raven and Prettyboy Reservoir watersheds, and in addition to intermittent fish barriers, the most commonly encountered habitat impairments that would limit fish migration and populations were low-flow (shallow depth) conditions and a lack of in-stream epifaunal vegetation and attached or fixed woody debris. Both of these conditions led to poor to very poor Physical Habitat Index values based on the Maryland Biological Stream Survey (MBSS) protocols (Kayzak, 2001) for about one-third of the stream reaches surveyed in the lower Loch Raven Reservoir

sub-watersheds, and about one-tenth of the reaches surveyed in the Prettyboy Reservoir sub-watersheds. On the other hand, fish barriers were more likely to restrict fish migration at sites surveyed in the Prettyboy Reservoir watershed than at sites surveyed in lower Loch Raven Reservoir watershed.

Additional information on the physical habitat conditions related to the biotic health of streams has been obtained through the MBSS. The MBSS surveys were conducted in the Baltimore reservoir watersheds in 1994, 1997, and 2000. Although the results of the MBSS surveys for the Loch Raven or Prettyboy Reservoir watersheds were not readily available at the time of this review, results for the Carroll County part of the Liberty Reservoir watershed were summarized as part of the watershed characterization (Maryland Department of Natural Resources, 2002a). In addition, the MBSS 2000 survey in the Liberty Reservoir watershed was augmented by MBSS Stream Waders, which nearly doubled the number of surveyed sites.

On the basis of the MBSS results, physical habitat conditions in the Liberty Reservoir watershed for most stream sites were rated fair to good. Only 5 of nearly 100 assessed sites were rated as poor and only 1 site as very poor. Based on habitat conditions, Liberty Reservoir watershed streams scored an average of 6.47 on a scale of 1 (worst) to 10 (best). For this size watershed, a score of 6.0 or less implies restoration is needed and a score of 8 or greater implies protection is recommended.

Two other measures of tributary biotic conditions in the Baltimore reservoir watersheds have been provided by routine benthic and fish surveys conducted as part of the MBSS. On the basis of surveys conducted in 1994, 1997, and 2000 in the Liberty Reservoir watershed, with the latter being augmented by additional data collection through the MBSS Stream Waders program, the Maryland Department of Natural Resources and Carroll County summarized survey findings as listed below (Maryland Department of Natural Resources and Carroll County, 2002a).

In relation to benthos integrity, the MBSS Program assessed 58 monitoring sites throughout the Liberty Reservoir watershed between 1995 and 2000. An additional 52 sites were sampled by citizen volunteers in 2000. Relative to reference streams, about 53 percent of the sites were considered good (or minimally degraded) with respect to reference stream conditions. A total of 16 sites (28 percent) were rated poor, with degraded conditions in relation to reference sites.

In relation to macro-invertebrate communities, Liberty Reservoir watershed streams scored an average of 6.89 on a scale of 1 (best) to 10 (worst). For nontidal watershed areas of this size, a score of less than 6 implies restoration is needed and a score of 8 implies protection is recommended.

In relation to fish communities, most streams in the Liberty Reservoir watershed were rated fair to good on the basis of MBSS data obtained between 1995 and 2000. The rating fair to good implies a generally diverse range of fish species are present at a site. Only a few sites were rated as poor.

In relation to fish communities, the Liberty Reservoir watershed streams generally are in good condition. The average site score of 8.87 on a scale of 1 (worst) to 10 (best) for streams in the Liberty Reservoir watershed implied that individual sites with scores of 8 or greater should be protected.

Although MBSS summaries were not available for the Loch Raven and Prettyboy Reservoir watersheds, these surveys form an important part of the Baltimore reservoir long-term monitoring strategy. The surveys provide the opportunity to periodically assess the biotic health of the reservoir watershed streams in a systematic and well-documented manner. Although surveys have been conducted since 2000 in all three reservoir watersheds, results have not been summarized. Ultimately, however, the MBSS data will provide for trend analysis related to the biotic health of reservoir streams.

Whereas the long-term monitoring program has helped identify subbasins that appeared to be sources of excessive nutrients and sediment, the collective components of the source-water assessments—watershed characterizations, short-term monitoring and synoptic surveys, stream-corridor and stability studies, and stream-habitat and biota surveys—enable the RWMA partners to describe the state of the watershed tributaries within these subbasins, identify impaired tributaries, develop restoration activities to address those impairments, and prioritize impaired streams for restoration. From these collective studies, however, it also is apparent that most impaired streams are located in headwater areas. Given their general location, the intensity of survey efforts to correctly identify the nature of stream impairments, and the resources required for reducing or eliminating impairments, it likely will take considerable time to restore impaired streams. Therefore, it is important to realize that it is likely to take considerable time before the full effects of restoration activities become apparent at the downstream tributary monitoring stations operated as part of the long-term monitoring program for major subbasins within each of the reservoir watersheds.

References Cited

Baltimore City Department of Public Works, 1996, Volume I: Selected graphics, report conclusions, and discussion: Baltimore, Maryland, Reservoir Watershed Management Progress Report, 128 p.

Baltimore City Department of Public Works, 2000, Appendix C: Reservoir water quality assessment for Loch Raven, Prettyboy and Liberty reservoirs, Interim Report, in Action report 2000 for the reservoir watersheds: Baltimore, Maryland, Watershed Protection Committee, 37 p.

Baltimore City Department of Public Works, 2001, Reservoir water quality assessment for Loch Raven, Prettyboy, and Liberty reservoirs: Baltimore, Maryland, 39 p.

Baltimore County Department of Environmental Protection and Resource Management, 1998, Gunpowder watershed characterization: Research design and monitoring procedures: Towson, Maryland, Baltimore County, 13 p.

Baltimore County Department of Environmental Protection and Resource Management, 2008, Prettyboy Reservoir watershed characterization report: Towson, Maryland, 101 p.

Frink, C.F., 1991, Estimating nutrient exports to estuaries: Journal of Environmental Quality, v. 20, no. 4, p. 717–724.

Interstate Commission on the Potomac River Basin, 2006, Modeling framework for simulating hydrodynamics and water quality in the Prettyboy and Loch Raven Reservoirs, Gunpowder River basin, Maryland: Rockville, Maryland, 230 p., plus appendixes.

Kayzak, P.F., 2001, Maryland Biological Stream Survey sampling manual: Annapolis, Maryland, Maryland Department of Natural Resources, 83 p.

Maryland Department of the Environment, 2003, Water quality and management plan for the Loch Raven watershed study area in Baltimore County, Maryland: Annapolis, Maryland, 180 p.

Maryland Department of the Environment, 2004, Source water assessment for Loch Raven watershed: Baltimore, Maryland, 99 p., plus appendixes.

Maryland Department of the Environment, 2006a, Nutrient synoptic survey in the Prettyboy watershed, Baltimore and Carroll Counties, Maryland, 2005, as part of a Watershed Restoration Action Strategy: Baltimore, Maryland, 20 p.

Maryland Department of the Environment, 2006b, Stream corridor assessment survey for the Prettyboy reservoir watershed: Baltimore, Maryland, 59 p.

Maryland Department of Natural Resources, 1998, Maryland clean water action plan, final (1998) report on unified watershed assessment, watershed prioritization and plans for restoration action strategies: Annapolis, Maryland, Clean Water Action Plan Technical Workgroup, 39 p., plus appendixes.

Maryland Department of Natural Resources, 2002a, Liberty reservoir watershed characterization-Carroll County portion, Maryland Department of Natural Resources: Annapolis, Maryland, and Carroll County, Maryland, 96 p.

Maryland Department of Natural Resources, 2002b, Nutrient and biological synoptic surveys in portions of the Carroll County drainage to Liberty reservoir: Annapolis, Maryland, 38 p.

Maryland Department of Natural Resources, 2002c, Liberty reservoir stream corridor assessment: Annapolis, Maryland, 103 p.

Winfield, G.L., and Sakai, J., 2003, Task 6: Final Liberty watershed assessment: Baltimore, Maryland, Department of Public Works, Bureau of Water and Wastewater, Project No. 812 On-Call Engineering Services, 110 p., plus appendix.

Appendix B: Descriptions of Data Collected at Watershed Tributary and Reservoir Monitoring Stations

Appendixes

Appendix B1. Baltimore City monitoring network station locations, 2007.

[N, north; W, west; NPDES, National Pollutant Discharge Elimination System; USGS, U.S. Geological Survey; WP, weather permitting, usually April through November or December]

Watershed	Station type	Reservoir location	Station identifier	Latitude, N (degrees, minutes, seconds)	Longitude, W (degrees, minutes, seconds)	Station location	Associated USGS gage	Comment
Liberty	Reservoir	Lower	NPA0042	39 23 20.00668	76 52 51.61888	Liberty Gatehouse, maximum depth 105 feet		
		Lower	NPA0059	39 23 48.47139	76 53 10.61397	Liberty at Route 26 Bridge, mid-channel		WP, requires boat
		Middle	NPA0067	39 25 04.09886	76 52 56.28529	Liberty at Oakland Road Point, near intake		WP, requires boat
		Upper	NPA0105	39 26 56.59918	76 52 40.35257	Liberty at Nicodemus/Deer Park Bridge, mid-channel		
	Tributary		BEA0015	39 29 22.55919	76 54 10.28399	Beaver Run at Hughes Road	01586210	
			LMR0015	39 25 36.29438	76 57 37.85655	Little Morgan Run at Bartholow Road		
			MDE0026	39 27 45.72575	76 54 27.22165	Middle Run at Louisville Road		
			MOR0040	39 27 07.22750	76 57 18.84300	Morgan Run at London Bridge Road	01586610	
			NPA0165	39 30 03.78299	76 53 00.61143	North Branch Patapsco at Route 91 (below outfall)	01586000	
			UZP0002	39 29 47.68237	76 52 12.78678	Bonds Run at Hollingsworth Road		
	NPDES		STP7704	39 30 14.34065	76 53 08.64558	Effluent Congoleum manufacturing plant		Grab, if discharge
			STP8000	39 35 14.32810	76 51 06.05665	Effluent Roy F. Weston manufacturing plant		Grab, if discharge
Loch Raven	Reservoir	Lower	GUN0142	39 25 50.86953	76 32 39.57712	Loch Raven Gatehouse		
		Middle	GUN0156	39 26 50.67932	76 33 20.68793	Loch Raven at Loch Raven Drive Bridge		Bridge access
		Middle	GUN0171	39 27 04.77082	76 34 08.44553	Loch Raven between picnic and golf course areas		WP, requires boat
		Upper	GUN0174	39 27 46.16274	76 34 49.59077	Loch Raven at Dulaney Valley Road Bridge		Bridge access
		Upper	GUN0190	39 29 00.28084	76 34 53.65176	Loch Raven at power lines		WP, requires boat
	Tributary		BEV0005	39 29 08.36173	76 38 44.59508	Beaver Dam Run at Beaver Run Lane	01583600	
			DVB0000	39 27 58.10206	76 32 43.51881	Dulaney Valley Branch at Loch Raven Drive		
			GUN0258	39 32 59.29999	76 38 09.95642	Gunpowder Falls at Glencoe Road	01582500	
			GUN0387	39 37 08.38463	76 41 26.10235	Gunpowder Falls at Falls Road	01581920	
			GUN0398	39 37 08.62153	76 42 23.89133	Gunpowder Falls 500 feet below Prettyboy Dam		
	(Pond)		JNR0003	39 28 02.09216	76 33 29.69286	Jenkins Run at Dulaney Valley Road		Grab, if flow
			LIT0002	39 36 07.34671	76 37 19.03368	Little Falls at Blue Mount Road	01582000	
			WGP0050	39 30 38.38894	76 40 36.83822	Western Run at Western Run Road	01583500	
	NPDES		STP8005	39 35 50.14921	76 50 03.52134	Effluent Hampstead Wastewater Treatment Plant		Grab, if discharge
Prettyboy	Reservoir	Lower	GUN0399	39 37 11.37820	76 42 26.02913	Prettyboy Dam Gatehouse		
		Lower	GUN0401	39 37 14.12232	76 42 36.07822	Prettyboy 1,000 feet upstream of Prettyboy Dam		WP, boat
		Middle	GUN0437	39 38 52.33959	76 45 19.25522	Prettyboy at Beckleysville Road Bridge		Bridge access
	Tributary		GOB0017	39 37 33.47745	76 46 22.06942	Georges Run at Georges Creek Road	01581870	
			GRG0013	39 39 16.99906	76 46 46.13531	Grave Run at Gunpowder Road	01581830	
			GUN0476	39 41 20.98079	76 46 49.77908	Gunpowder Falls at Gunpowder Road	01581810	
	NPDES		STP8006	39 39 27.88889	76 52 46.84074	Effluent Manchester Wastewater Treatment Plant		Grab, if discharge

Appendix B2. Baltimore City monitoring for Liberty Reservoir: treatment facility, reservoir, and tributaries—sample location, water-quality parameters, and frequency.

[NPDES, National Pollutant Discharge Elimination System; TF, treatment facility, R, raw, or T, treated, water sample or measurement, taken in facility; raw-water sample or measurement taken at all other locations; D, daily; W, weekly; M, monthly; 2xW, twice weekly; 2xM, twice monthly; (S), spring-summer-fall (approximately April through November); (W), winter (approximately December through March); E, storm event; *, permitted discharge, grab sample, if flow; ---, not sampled; C, degrees Celsius, mg/L, milligrams per liter, µg/L micrograms per liter; L, liter; mL, milliliter; #, number; MP, most probable; µmhos/cm, micromhos per centimeter; CaCO₃, calcium carbonate]

Water-quality parameter	Ashburton		Liberty Reservoir — Reservoir site identifier				Watershed tributary site identifier						NPDES site identifier	
	TF, Raw	TF, Treated[1]	NPA0042	NPA0059	NPA0067	NPA0105	BEA 0015	LMR 0015	MDE 0026	MOR 0040	NPA 0165	UZP 0002	STP7704*	STP8000*
Temperature, air, C	---	---	2xM(S), M(W)	2xM(S)	2xM(S)	2xM(S), M(W)	M	M	M	M	M	M	M	M
Temperature, water, C	D	D	2xM(S), M(W)	2xM(S)	2xM(S)	2xM(S), M(W)	M	M	M	M	M	M	M	M
Color (True), color units	2xW	2xW	2xM(S), M(W)	---	---	---	---	---	M	---	---	---	---	---
Secchi disc, meters	---	---	2xM(S), M(W)	2xM(S)	---	2xM(S), M(W)	---	---	---	---	---	---	---	---
Turbidity, nephelometric units	D	D	2xM(S), M(W)	M(S)	M(S)	M	M	M	M	M	M	M	M	M
Total solids, mg/L	M	M	---	---	---	M	M	M	M	M	M	M	M	M
Total suspended solids, mg/L	---	---	---	---	---	---	M, E	M	M	M, E	M, E	M	M	M
Volatile suspended solids, mg/L	---	---	---	---	---	---	E only	---	---	E only	E only	---	---	---
Chlorophyll-a, µg/L	---	---	2xM(S), M(W)	2xM(S)	2xM(S)	2xM(S), M(W)	---	---	---	---	---	---	---	---
Total algal count, #/mL	W	W	2xM(S), M(W)	M(S)	M(S)	M	---	---	---	---	---	---	---	---
Algal identification	W	W	2xM(S), M(W)	M(S)	M(S)	M	---	---	---	---	---	---	---	---
Total and fecal coliforms, MP#/100 mL	D	D	---	---	---	---	---	---	---	---	---	---	---	---
Cryptospiridium and Giardia, #/L	M	---	M	---	---	M	---	---	---	---	---	---	---	---
Specific conductance, µmhos/cm	M	M	2xM(S), M(W)	2xM(S)	2xM(S)	2xM(S), M(W)	M	M	M	M	M	M	M	M
Dissolved solids, mg/L	---	M	2xM(S), M(W)	M(S)	M(S)	M	---	M	M	M	M	M	M	M
Hardness, mg/L as CaCO₃	M	D	M	M(S)	M(S)	M	M	M	M	M	M	M	M	M
pH, Standard units	D	D	2xM(S), M(W)	M(S)	M(S)	M	M	M	M	M	M	M	M	M
Alkalinity, mg/L as CaCO₃	D	D	2xM(S), M(W)	M(S)	M(S)	M	M	M	M	M	M	M	M	M
Dissolved oxygen, mg/L	2xW	2xW	2xM(S), M(W)	2xM(S)	2xM(S)	2xM(S), M(W)	M	M	M	M	M	M	M	M
Manganese, mg/L	D	D	2xM(S), M(W)	M(S)	M(S)	M	---	---	M	---	---	---	---	---
Iron, mg/L	3xW	3xW	M	M(S)	M(S)	M	---	---	---	---	---	---	---	---
Phosphorus, total, mg/L	M	M	2xM(S), M(W)	M(S)	M(S)	M	M, E	M	M	M, E	M, E	M	M	M
Nitrogen, nitrate, mg/L	M	M	M	M(S)	M(S)	M	M, E	M	M	M, E	M, E	M	M	M
Nitrogen, ammonium, mg/L	M	M	M	M(S)	M(S)	M	M	M	M	M	M	M	M	M
Nitrogen, total Kjeldahl, mg/L	---	---	---	---	---	---	E only	---	---	E only	E only	---	---	---
Carbon, total organic, mg/L	2xM	2xM	---	---	---	---	---	---	---	---	---	---	---	---
Trihalomethanes, µg/L	M	M	---	---	---	---	---	---	---	---	---	---	---	---
Haloacetic acids, µg/L	M	M	---	---	---	---	---	---	---	---	---	---	---	---
Sodium, mg/L	M	M	---	---	---	---	M	M	M	M	M	M	M	M
Chloride, mg/L	M	M	2xM(S), M(W)	M(S)	M(S)	M	M	M	M	M	M	M	M	M

[1] Additional constituents only for treated water: calcium, magnesium, potassium, aluminum, arsenic, sulfate, silica, fluoride, residual chlorine, and selected metals, radionuclides, and synthetic and volatile organic compounds

Appendix B3. Baltimore City monitoring for Loch Raven Reservoir: treatment facility, reservoir, and tributaries—sample location, water-quality parmeters, and frequency.

[NPDES, National Pollutant Discharge Elimination System; TF, treatment facility, R, raw, or T, treated, water sample or measurement, taken in facility; raw-water sample or measurement taken at all other locations; D, daily; W, weekly; M, monthly; 2xM, twice weekly; 2xW, twice weekly; 2xM, twice monthly; (S), spring-summer-fall (approximately April through November); (W), winter (approximately December through March); E, storm event; *, permitted discharge, grab sample, if flow; **, grab sample, if flow; ---, not sampled; C, degrees Celsius; mg/L, milligrams per liter; µg/L, micrograms per liter; L, liter; mL, milliliter; #, number; MP, most probable; µmhos/cm, micromhos per centimeter; CaCO₃, calcium carbonate]

Loch Raven Reservoir

Water-quality parameters	Montebello TF, Raw	Montebello TF, Treated[1]	GUN 0142	GUN 0156	GUN 0171	GUN 0174	GUN 0190	BEV 0005	DVB 0000	GUN 0387	GUN 0256	WGP 0050	LIT 0002	GUN 0398*	JNR 0003*	STP 8005*,**,***
			Reservoir site identifier					Watershed tributary site identifier						NPDES site identifier		
Temperature, air, C	D	D	2xM(S), M(W)	M(W)	2xM(S)	M(W)	2xM(S)	M	M	2xM	M	M	M	M	M	M
Temperature, water, C	D	D	2xM(S), M(W)	M(W)	2xM(S)	M(W)	2xM(S)	M	M	2xM	M	M	M	M	M	M
Color (True), color (apparent) units	D	D	2xM(S), M(W)	M(W)	M(S)	M(W)	M(S)	---	---	---	---	---	---	M	---	---
Secchi disc, meters	---	---	2xM(S), M(W)	M(W)	2xM(S)	M(W)	2xM(S)	---	---	---	---	---	---	---	---	---
Turbidity, nephelometric units	D	D	2xM(S), M(W)	M(W)	M(S)	M(W)	M(S)	M	M	M	M	M	M	M	M	M
Total solids, mg/L	M	M	---	---	---	---	---	M	M	M	M	M	M	---	M	M
Total suspended solids, mg/L	---	---	---	---	---	---	---	M, E	M	M	M, E	M	M	---	M	M
Volatile suspended solids, mg/L	---	---	---	---	---	---	---	E	---	---	E	E	---	---	---	---
Chlorophyll-α, µg/L	---	---	2xM(S), M(W)	M(W)	2xM(S)	M(W)	2xM(S)	---	---	---	---	---	---	---	---	---
Total algal count, #/mL	W	W	2xM(S), M(W)	---	---	---	---	---	---	---	---	---	---	---	---	---
Algal identification	W	W	2xM(S), M(W)	---	---	---	---	---	---	---	---	---	---	---	---	---
Total and fecal coliforms, MP#/100 mL	D	D	---	---	---	---	---	---	---	---	---	---	---	---	---	---
Cryptospiridium and Giardia, #/L	M	M	---	---	---	---	---	---	---	---	---	---	---	---	---	---
Specific conductance, µmhos/cm	M	M	2xM(S), M(W)	M(W)	2xM(S)	M(W)	2xM(S)	M	M	2xM	M	M	M	M	M	M
Dissolved solids, mg/L	M	M	2xM(S), M(W)	M(W)	M(S)	M(W)	M(S)	M	M	M	M	M	M	M	M	M
Hardness, mg/L as CaCO₃	D	D	2xM(S), M(W)	M(W)	2xM(S)	M(W)	2xM(S)	M	M	2xM	M	M	M	M	M	M
pH, Standard units	D	D	2xM(S), M(W)	M(W)	2xM(S)	M(W)	2xM(S)	M, E	M	M	M, E	M, E	M	M	M	M
Alkalinity, mg/L as CaCO₃	D	D	2xM(S), M(W)	M(W)	M(S)	M(W)	M(S)	M, E	M	M	M, E	M, E	M	M	M	M
Dissolved oxygen, mg/L	D	D2	2xM(S), M(W)	M(W)	2xM(S)	M(W)	2xM(S)	M	M	2xM	M	M	M	---	M	M
Manganese, mg/L	D	D	2xM(S), M(W)	M(W)	M(S)	M(W)	M(S)	---	---	---	---	---	---	M	---	---
Iron, mg/L	3xW	3xW	---	---	---	---	---	---	---	---	---	---	---	M	---	---
Phosphorus, total, mg/L	M	M	2xM(S), M(W)	M(W)	M(S)	M(W)	M(S)	M, E	M	M	M, E	M, E	M	M	M	M
Nitrogen, nitrate, mg/L	M	M	2xM(S), M(W)	M(W)	M(S)	M(W)	M(S)	M, E	M	M	M, E	M, E	M	M	M	M
Nitrogen, ammonium, mg/L	M	M	2xM(S), M(W)	M(W)	M(S)	M(W)	M(S)	M	M	M	M	M	M	M	M	M
Nitrogen, total Kjeldahl, mg/L	---	---	---	---	---	---	---	E	---	---	E	E	---	---	---	---
Carbon, total organic, mg/L	2xM	2xM	---	---	---	---	---	---	---	---	---	---	---	---	---	---
Trihalomethanes, µg/L	M	M	---	---	---	---	---	---	---	---	---	---	---	---	---	---
Haloacetic acids, µg/L	M	M	---	---	---	---	---	---	---	---	---	---	---	---	---	---
Sodium, mg/L	M	M	---	---	---	---	---	M	M	M	M	M	M	---	M	---
Chloride, mg/L	M	M	---	---	---	---	---	M	M	M	M	M	M	---	M	M

[1] Additional constituents for treated water: calcium, magnesium, potassium, aluminum, arsenic, sulfate, silica, fluoride, residual chlorine, and selected metals, radionuclides, and synthetic and volatile organic compounds

Appendix B4. Baltimore City monitoring for Prettyboy Reservoir: reservoir and tributaries—sample location, water-quality parameters, and frequency.

[NPDES, National Pollutant Discharge Elimination System; M, monthly; (S), spring-summer-fall (approximately April through November); (W), winter (approximately December through March); **, permitted discharge, grab sample, if flow; --, not sampled; C, degrees Celsius; mg/L, milligrams per liter; μg/L, micrograms per liter; L, liter; mL, milliliter; #, number; MP, most probable; μmhos/cm, micromhos per centimeter; CaCO₃, calcium carbonate]

Water-quality parameters	Reservoir site identifier			Prettyboy Reservoir — Watershed tributary site identifier			NPDES site identifier
	GUN0399	GUN0401	GUN0437	GOB0017	GRG0013	GUN0476	STP8006**
Temperature, air, C	M(W)	M(S)	M	M	M	M	M
Temperature, water, C	M(W)	M(S)	M	M	M	M	M
Color (True), color units	M(W)	M(S)	M	--	--	--	--
Secchi disc, meters	M(W)	M(S)	M	--	--	--	--
Turbidity, nephelometric units	M(W)	M(S)	M	M	M	M	M
Total solids, mg/L	--	--	--	M	M	M	M
Total suspended solids, mg/L	--	--	--	M	M	M	M
Volatile suspended solids, mg/L	M(W)	M(S)	M	--	--	--	--
Chlorophyll-a, μg/L	M(W)	M(S)	--	--	--	--	--
Total algal count, #/mL	M(W)	M(S)	--	--	--	--	--
Algal identification	--	--	--	--	--	--	--
Total and fecal coliforms, MP#/100 mL	--	--	--	--	--	--	--
Cryptospiridium and Giardia, #/L	M(W)	M(S)	M	M	M	M	M
Specific conductance, μmhos/cm	--	--	--	M	M	M	M
Dissolved solids, mg/L	--	--	--	--	--	--	--
Hardness, mg/L as CaCO₃	M(W)	M(S)	M	M	M	M	M
pH, Standard units	M(W)	M(S)	M	M	M	M	M
Alkalinity, mg/L as CaCO₃	M(W)	M(S)	M	M	M	M	M
Dissolved oxygen, mg/L	M(W)	M(S)	M	--	--	--	--
Manganese, mg/L	--	--	--	--	--	--	--
Iron, mg/L	M(W)	M(S)	M	M	M	M	M
Phosphorus, total, mg/L	--	--	--	--	--	--	--
Nitrogen, nitrate, mg/L	M(W)	M(S)	M	M	M	M	M
Nitrogen, ammonium, mg/L	M(W)	M(S)	M	M	M	M	M
Nitrogen, total Kjeldahl, mg/L	--	--	--	--	--	--	--
Carbon, total organic, mg/L	--	--	--	--	--	--	--
Trihalomethanes, μg/L	--	--	--	--	--	--	--
Haloacetic acids, μg/L	--	--	--	--	--	--	--
Sodium, mg/L	--	--	--	--	--	--	--
Chloride, mg/L	--	--	--	M	M	M	M

Appendix C: Review of Baltimore Reservoir Ashburton and Montebello Treatment Facilities Laboratory Quality-Assurance Plans

Contents

Introduction

The following quality-assurance (QA) plans for the Ashburton and Montebello treatment facility laboratories were requested and reviewed as part of the retrospective review, and are included in this Appendix. Except for minor reformatting, the plans are presented as they were received. For selected headings, no additional text appears in plan after the heading.

Ashburton Laboratory Quality Assurance Plan, Revision 3, 2007

This plan was obtained in March 2007 from Savita Bagel, Laboratory Manager, Ashburton Laboratory, Baltimore, Maryland. It has been minimally reformatted for inclusion in this Appendix.

1. **Organization Chart, Line Authority - see list (attached list not requested)**

2.

3. **Analytical Procedures (see the Standard Operating Procedures (SOP) manual)**

4. **Sample Handling Procedures**

   Sample Acceptance and Logging—There is a sample log for all samples coming into the laboratory. All samples must be collected, stored and preserved in accordance with EPA guidelines. More specific instructions are in the SOPs.

   Sample Rejection—Samples are rejected for improper labeling, collection, storage or holding times.

   Sample Disposable—Samples are disposed of after all analyses are completed or at the end of the holding time.

   Sample Storage—Chlorine and pH analyses must be done immediately when the sample comes into the lab. For most other analyses the samples are stored in the refrigerators before analysis. For metals analyses the samples should be preserved with nitric acid to less than pH 2.

   Sample Tracking—All samples should be recorded in the sample log book.

   Chain of Custody—Is needed for samples being transported to or from Montebello, including samples for metals analysis.

5. **Sampling Procedures**

   Containers—The container must be appropriate for the intended analysis and must be labeled with the location, date and time of collection. Sterile bottles with sodium thiosulfate are used for micro samples. Acid-washed bottles are needed for metal analyses.

   Preservation—All, except samples for metal analysis, must be kept on ice or refrigerated until analyzed. If analyses for nitrate, ammonia and phosphates cannot be completed within 48 hours, the samples are preserved by acidifying to < pH 2 with concentrated sulfuric acid. Samples for metal analysis should be acidified to < pH 2 with concentrated nitric acid. They must be held for 16 hours after acidification and then can be held for up to 6 months. If the turbidity is > 1 NTU, the sample must be digested. (see SOPs.)

   QC Samples—Should be done quarterly for fluoride, nitrites, and nitrates and as many other parameters as possible. The complete analysis should be done on the yearly Performance Evaluation samples each spring. As many analysts as possible should complete each analysis.

   Documentation—All samples must be logged into the sample log book and/or have a paper form with the required information. Included must be the name of the sampler, date and location of the sample and a list of parameters for analysis.

   Special Instructions—Care must be exercised to take samples that will be representative of the water being tested and to avoid contamination of the sample at the time of collection or in the period before analysis.

   Plant Process Samples—Samples can be taken from the sink taps, anytime, except when the water flow thru the plant has been changed recently. A change in flow may affect the water quality temporarily. Samples should be analyzed immediately when possible.

Metals Analyses Samples—The tap should be opened and the water allowed to run to waste for 2 to 3 minutes or for a sufficient time to permit clearing of the service line. Chlorine and pH determinations must be done at this time. The flow from the tap should then be restricted to one that will permit filling the bottle without splashing. The bottles can be filled almost to the top leaving enough air space to permit mixing. For a first draw sample for lead analysis, the line should be thoroughly flushed and then allowed to sit unused for 6 to 8 hours. The sample should then be collected as soon as the tap is opened.

Utility Maintenance Samples—There are no restrictions on these samples except that we need to know the location and the time collected. These analyses are not done by approved methods and are only an approximation, but good enough to tell whether city water, sewage or ground water is involved.

Watershed Samples—Are collected by the watershed samplers.

Water Quality Management Samples—Storm water runoff samples for nutrient analyses are preserved before they come to the lab.

Distribution Samples—Must be collected by a State Certified Sampler and are almost always analyzed at Montebello.

Waste Lake Samples—All composite samples must be analyzed each week to meet the requirements of the NPDES permit for the plant.

6. Calibration Procedures

Standards Source—ERA, NSI, SPEX, Fisher, Perkin Elmer.

Comparability Checks—New standards are run against old standards with a QC sample.

Frequency—The pH meter is calibrated each morning and afternoon with three certified buffers. The calibration is checked with each use. The balance calibration is checked monthly. All turbidimeters in the plant are calibrated each month by the Instrumentation group. Calibration of other instruments is done each day of use. A set of at least 3 standards and a blank must be used. An appropriate standard or QC sample must be checked after a set number (usually 10) of samples. Standards and QC samples must be run again at the end of the run. More specific instructions are in each SOP.

7. Documentation

There are calibration books for pH and fluoride as well as for the digital berets and balances. For other analyses the analyst keeps the calibration documentation with the sample results.

8. Data Reduction, Validation and Reporting

Calculations

Units Units must be clearly marked as to mg/L or (text ends here)

Transcription/Transfer Each analyst records their results on the report form.

Report Format The report format varies with the type of sample.

Documentation

9. QC Checks—see SOPs for specific instructions.

Reagent Blanks—Are done for each set of analyses.

Replicate Analyses—At least 10 % of samples must be duplicated.

Check Sample Recoveries

Matrix Spike Recoveries—Are done for every sample analyzed by graphite furnace. For fluorides, the spikes are done each day. For nitrates, the spikes should be done on at least 10 % of the samples or whenever the matrix changes.

Instrument Control Standard Response

Internal Standard Response

Control Charts—Are done for fluoride, nitrates, nitrites and all metal analyses.

Documentation—If the QC is not acceptable, the samples must be run again. Any out of control sample or standard on the control charts must be documented with the corrective action: The sample could be rerun or recalibration was done or maintenance of the instrument was needed.

10. Specific Routine Procedures

Accuracy—All graphite furnace samples are spiked. 10% of other samples are spiked.

Precision—Duplicates are done to demonstrate precision.

Completeness

Timeliness—All samples must be analyzed within the approved holding time unless clearly marked on the report of results. Compliance samples must always be done within the approved holding time.

Legibility

Clarity

11. Schedules of internal and external system and data quality audits and inter laboratory comparisons.

The EPA performance evaluation samples for chemistry are done each year for all analyses that are done in the lab. At least once each quarter, QC samples are done for nitrates, nitrites and fluorides. QC samples are included for each of the metals analyses each time they are run.

12. Preventive Maintenance

Operating Manuals—There is a file drawer for operating manuals and instruction books for most equipment. A file folder of instructions and a record of repairs and replacement parts will be included there.

Service Schedule—Both balances are cleaned and calibrated each year by American Scale. Class S weights are used to check the balance calibration each month.

Spare Parts Inventory—Some spare parts are kept on hand including parts for the glass still and various bulbs for the spectrophotometer and the turbidimeter. Backup equipment is available for the ion meter, the pH meter, the turbidimeter and the spectrophotometer. There is cooperation with the Montebello Lab for emergencies.

Service Agreements—The annual service contract agreement for the Perkin Elmer Atomic absorption spectrometer includes two preventive maintenance visits each year.

Documentation—All service and repairs should be documented in the appropriate file folder in the equipment file. Records for the AA are in the desk nearest the AA.

13. Corrective Action

QC Failure—If the QC results are not acceptable, the analysis must be repeated.

PE Failure—Every aspect of the analysis must be examined to determine the problem. Correction must be as soon as possible and steps taken to ensure that the problem will not occur again.

Audit Deficiency—Must be corrected as soon as possible.

Complaint

14. Record Keeping Procedures

Keep original data for at least 5 years. Monthly reports are kept on disk (two copies) and at least one paper copy is in the lab as well. As needed, reports are kept on disk as well as on paper. Reports are distributed as needed to other people.

Montebello Laboratory Quality Assurance Plan for Chemical Analysis, Revision 3, 2007

This plan was obtained in March 2007 from Lisa Jones, Laboratory Manager, Montebello Laboratory, Baltimore, Maryland. It has been minimally reformatted for inclusion in this Appendix.

1. Organization Chart

INEZ HAWK, LABORATORY TECHNICAL ADMINISTRATOR

LISA JONES, LABORATORY TECHNICAL SUPERVISOR

DEBORAH PITTS, MICROBIOLOGIST SUPERVISOR

JOSEPH BRENNAN, CHEMIST II

OMACHILE TAUPYEN, CHEMIST I

MARIA REED, MICROBIOLOGIST II

KAREN CAMPBELL, LAB ASSISTANT II

JOHN HOHMAN, POLLUTION CONTROL ANALYST II

RICHARD NUSS, CHEMIST III

THO NGUYEN, CHEMIST III

All of the chemists have at least 20–30 hours of college level chemistry courses. Each new analyst is trained and checked by a senior analyst. The laboratory supervisor is responsible for implementing the QA plan. Each analyst is responsible for doing analyses with the required quality control.

The laboratory supervisor is responsible for making sure that all personnel are updated on changes in regulations and methodology. The analysts should be able to make the necessary changes with minimal assistance.

The Microbiologists have a Microbiological Quality Control and Procedures Manual. The Microbiologist Supervisor makes changes in methods and procedures. Now, while that position is vacant, the laboratory supervisor will make changes when necessary.

On weekends and holidays, one analyst does both the routine chemistry and microbiology. Therefore, all chemists and microbiologists must demonstrate the ability to perform the routine chemical and microbiological analyses that are done on weekends and holidays. This is done before they work a weekend.

2. Data Quality Objectives:

Compliance samples are done with approved methods and all appropriate quality control, being careful to observe required holding times. Plant process samples occasionally can be done more informally; e.g., manganese. Waste lake samples need to be done as quickly as possible with screening tests that do not necessarily require all the usual quality control.

3. Analytical Procedures-Dates of Revision (see the SOP manual)

Alkalinity-titration method-2/28/97

Ammonia-Electrode method-11/9/98

Calcium Carbonate Stability-2/11/97

Carbon dioxide-titrimetric method-12/20/95

Chloride-argentometric method-12/95

Chloride-see ion chromatography method-11/5/98

Chlorine-amperometric titration-2/27/97

Color-Visual comparison-11/6/98

Dissolved Oxygen-electrode-12/95

Fluoride-ion selective method-10/95

Hardness-EDTA titration-2/28/97

Iron-Phenanthroline method-12/95

Ion Chromatography for nitrate, chloride and sulfate-11/5/98

Jar test procedure-3/22/96

Manganese for plant process-11/6/98

Nitrate-electrode method-11/4/98

Nitrate-see ion chromatography method-11/5/98

Nitrite-colorimetric method-3/16/98

PH-electrometric-11/19/96

Phosphate-total-ascorbic acid method-11/16/95

Silica-molybdosilicate-12/95

Sulfates-see ion chromatography method-11/5/98

Threshold odor number (TON)-12/19/95

Total Organic Carbons 4/2000

Total suspended solids-Dried at 180°C-4/2002

Total Solids-11/3/98

Turbidity-Nephelometric Method-3/20/97

Volatile solids-11/3/98

Organics:

Trihalomethanes	EPA 524.2
HAA's	EPA 552
VOC's	EPA 524.2
EDB, DBCP	EPA 504
Organohalide Pesticides	EPA 505
Chlorinated Pesticides	EPA 508, 515.1
TTHM Formation Potential	EPA 510.1

4. Sampling Procedures

Containers—The container must be appropriate for the intended analysis and must be labeled with the location, date, time of collection and collector. Sterile bottles with sodium thiosulfate are used for micro samples. Acid-washed (25% HNO_3) bottles are needed for metal analyses. Leak samples are collected in clean glass pint bottles. Watershed samples are collected in acid (50% HCl) rinsed plastic liter bottles.

Preservation—All samples, except samples for metal analysis, must be kept on ice or refrigerated until analyzed. If analyses for nitrate, ammonia and phosphates cannot be completed within 48 hours; acidifying to < pH with concentrated sulfuric acid preserves the samples.

Samples for metal analysis should be acidified to pH 2 with concentrated trace metal grade nitric acid. They must be held for 16 hours after acidification and then can be held for up to 6 months.

Special Instructions—Care must be exercised to take samples that will be representative of the water being tested and to avoid contamination of the sample at the time of collection or in the period before analysis. Paperwork must be filled out in ink.

Plant Process Samples—Samples can be taken from the sink taps, anytime, except when the water flow through the plant has been changed recently. A change in flow may affect the water quality temporarily. Samples should be analyzed immediately when possible. Chlorine and pH must be done immediately. Immediately is considered to be within 15 minutes.

Metals Analyses Samples—The tap should be opened and the water allowed to run to waste for 2 to 3 minutes or for a sufficient time to permit clearing of the service line. Chlorine and pH determinations must be done at this time. The flow from the tap should then be restricted to one that will permit, filling the bottle without splashing. The bobbles can be filled almost to the top leaving enough air space to permit mixing. For a first draw sample for lead analysis, the line should be thoroughly flushed and then allowed to sit unused for 6 to 8 hours. The sample should then be collected as soon as the tap is opened.

Watershed Samples—Are collected by the watershed samplers and will come with all needed paperwork.

Organic Bottles—Must be cleaned according to the appropriate EPA protocol for each method.

Distribution Samples—Must be collected by a Certified Sampler and are always analyzed at Montebello.

5. Sample Handling Procedures

Sample Acceptance and Logging

All samples must be recorded in the sample log book and/or have a paper form with the required information. Included must be the name of the sampler, date and location of the sample and a list of parameters for analysis. All samples must be collected, stored and preserved in accordance with EPA guidelines. More specific instructions are in the SOPs. Samples from the Ashburton Lab should have a chain of custody with the appropriate information.

Sample Rejection—Samples are rejected for improper labeling, collection, storage or holding times.

Sample Storage—Chlorine and pH analyses must be done immediately when the sample comes into the lab. This means they must be done within 15 minutes of collection. For most other analyses the samples are stored in the refrigerators before analysis. For metals analyses the samples are to be preserved with nitric acid to less than pH 2 and then can be held at room temperature.

Sample Disposal—Samples are disposed of after all analyses are completed or at the end of the holding time.

Sample Tracking—All samples are to be recorded in the sample logbook. Microbiological samples for coliform analysis must be logged into the micro book and stored in the Micro refrigerator on the shelf for coliform samples.

Chain of Custody—Is needed for samples being transported to or from Ashburton, including samples for metals analysis.

6. Calibration Procedures

Standards Source ERA, SPEX, Fisher

Comparability Checks New standards are run against old standards with a QC sample.

Frequency The pH meter is calibrated each morning with two certified buffers. The calibration is checked with each use. The balance calibration is checked monthly. All turbidmeters in the plant are calibrated quarterly by the Instrumentation group. Calibration of other instruments is done each day of use. A set of at least 3 standards and a blank must be used. An appropriate standard or QC sample must be check after a set number (usually 10) of samples. Standards and quality control samples must be run again at the end of the run. More specific instructions are in each SOP.

Documentation There are calibration books for pH and fluoride and balances. For other analyses the analyst keeps the calibration documentation with the samples results.

7. Analytical Procedures

Standard Operating Procedures are in SOP manual and should be reviewed by the analysts frequently. The complete method citation is included at the beginning of the SOP. The date of the last revision is at the bottom of the SOP.

8. Data Reduction, Validation and Reporting

Units Units must be clearly marked.

Transcription/Transfer Each analyst records his or her results on the report form.

Report Format The report format varies with the type of sample.

Documentation results can only be reported if all the QC has been satisfactory.

9. QC Checks see SOPs for specific instructions.

Reagent Blanks are done for each set of analyses.

Replicate Analyses at least 10% of samples must be duplicated.

Check Sample Recoveries The required percentage range varies with the type of analyses.

Matrix Spike Recoveries For fluorides, the spikes are done each day. For nitrates the spikes should be done on at least 10% of the samples or whenever the matrix changes.

Instrument Control Standard Response varies for each instrument.

Control Charts are done for fluoride, nitrates, and nitrites.

Documentation If the QC is not acceptable, the samples must be run again. Any out of control recovery on the control charts must be documented with the corrective action:

The sample was rerun or recalibration was done or maintenance of the instrument was needed.

Accuracy 10% of all samples are spiked for certified analyses.

Timeliness All samples must be analyzed within the approved holding time unless results are clearly marked on the report. Compliance samples must always be done within the approved holding time.

Method Detection Limits (MDL) must be done at least annually by each new analyst.

Quality Control procedures for microbiology are included in the Microbiology Manual.

10. Schedules of internal and external system and data quality audits and inter-laboratory comparisons.

The EPA Performance Evaluation samples for chemistry are done each year for all analyses that are done in the lab.

As many analysts as possible should complete each analysis. At least once each quarter, QC samples are done for nitrates, nitrites and fluorides. QC samples are included for each of the ORGANICS analyses each time they are run.

11. Preventive Maintenance

Operating Manuals There is a file drawer for operating manuals and instruction books for most equipment. A file folder of instructions and a record of repairs and replacement parts for each instrument are included there.

Service Schedule All balances are cleaned and calibrated each year by American Scale. Class 1 weights are used to check the analytical balance calibration each month.

Spare Parts Inventory Some spare parts are kept on hand including parts for the glass still and various bulbs for the spectrophotometer and the turbidimeter. Backup equipment is available for the ion meter, the pH meter, the turbidimeter and the spectrophotometer. There is cooperation with the Ashburton Lab for emergencies.

Service Agreements The Ion Chromatograph is covered by a service contract with preventive maintenance visits. The same applies to the autoclave, dishwasher, and the Total Organic Carbon Analyzer. Service contracts are generally in effect for the GC/MS.

All service and repairs should be documented in the appropriate maintenance log book near the equipment.

12. Corrective Action

QC Failure- If the QC results are not acceptable, the analysis must be repeated. If the holding time has expired, the sample cannot be rerun and the results must re-check the SOP to see if everything was done correctly. Check to be sure that all reagents are correct. Depending on the type of analysis, recalibration might be required. Dilution of the sample might be helpful. Check with the laboratory supervisor or chemist III for more suggestions.

PE Failure every aspect of the analysis must be examined to determine the problem. Corrective action must be taken as soon as possible to ensure that the problem will not occur again. Check the SOP. Reagents should be checked. The instrument used may need to be serviced. Preventive maintenance procedures should be reviewed. The results of the investigation should be written up and given to the laboratory supervisor to be sent in the response to the State.

Audit Deficiency must be corrected as soon as possible.

13. Record Keeping Procedures

Original data in workbooks is kept for at least five years. Copies of the weekly Watershed data are sent to the watershed section.

Plant monthly reports are kept on disk (two copies) and at least one paper copy is in the lab as well. Copies are sent to the State MDE, to the Water System Manager, Water Systems Assistant Manager, the Water Quality Laboratory Administrator and the Water Quality Lab Supervisors.

Microbiological results are reported by telephone. The paperwork is filed for future reference.

Appendix D: Plant Ecology Group (PEG) Model of Seasonal Succession of Plankton in Freshwater

The Plankton Ecology Group (PEG) model sequentially describes the general trend of a spring bloom of small diatoms, followed by the progression during summer from large colonial green algae to large diatoms, then large dinoflagellates and (or) finally blue-green algae (Sommer and others, 1986). In so doing, the PEG model incorporates the relative importance of physical factors, nutrients (nitrogen and phosphorus), and grazing in shaping phytoplankton community structure throughout the growing season in freshwater lakes, as follows (from Sommer and others, 1986, with modified formatting):

a) Towards the end of winter, nutrient availability and increased light permit unlimited growth of the phytoplankton. A small crop of fast-growing algae, for example golden-brown (*Cryptophyceae sp.*) and centric diatoms develops.

b) This crop of small algae is grazed upon by herbivorous zooplanktonic species, which become abundant due to hatching from resting eggs and to high fecundity by high levels of edible algae.

c) Planktonic herbivores with short generation duration times increase their populations first and are followed by slower growing species.

d) The herbivore populations increase exponentially up to the point at which their density is high enough to produce a community filtration rate, and so cropping rate, which exceeds the reproduction rate of the phytoplankton.

e) As a consequence of herbivore grazing, the phytoplankton biomass decreases rapidly to very low levels.

f) There then follows a 'clear-water' equilibrium phase which persists until inedible algae species develop in significant numbers. Nutrients are re-cycled by the grazing process and can accumulate during the 'clearwater' phase.

g) Herbivorous zooplanktonic species become food-limited and both their body weight per unit length and their fecundity declines. This results in a decrease in their population densities and biomasses.

h) Fish predation accelerates the decline of herbivorous planktonic populations to very low levels and this trend is accompanied by a shift towards a smaller average body size amongst the surviving crustaceans.

i) Under the conditions of reduced grazing pressure and sustained non-limiting concentrations of nutrients, the phytoplankton summer crops start to build up. The composition of the phytoplankton becomes complex due to both the increase in species richness and to the functional diversification into small 'undergrowth' species, which are available as food for filter-feeders, and into large 'canopy' species, which are only consumed by specialists such as raptors or parasites.

j) At first, the edible algae (such as golden-brown (*Cryptophyceae sp.*) and inedible colonial green algae become predominant. They deplete the soluble reactive phosphorus to nearly undetectable levels.

k) From this time onwards, the algal growth becomes nutrient-limited and this prevents an explosive growth of 'edible' algae. Grazing by predator-controlled herbivores balances the nutrient-limited growth rate of edible algal species.

l) Competition for phosphate leads to a replacement of green algae by large diatoms, which are only partly available to zooplankton as food.

m) Silica-depletion leads to a replacement of the large diatoms by large dinoflagellates and/or blue green algae (*Cyanophyta sp.*).

n) Nitrogen depletion ultimately favors a shift to nitrogen-fixing species of filamentous blue-green algae.

o) Larger species of crustacean herbivores are replaced by smaller species and by rotifers. These small species are less vulnerable to fish predation and are less affected by interference with their food collecting apparatus which can be caused by some forms of inedible algae. Accordingly, their population mortality is lower and their fecundity is higher than that of the larger species.

p) The smaller species of herbivores coexist under a persistent fish predation pressure and the increased possibility of food partitioning, which is associated with the greater species complexity of the phytoplankton.

q) The population densities and species composition of the zooplankton fluctuate throughout the summer, the latter being also influenced by temperature.

r) The period of autogenic succession is terminated by factors related to physical changes, which includes increased mixing depth resulting in nutrient replenishment and a deterioration of the effective underwater light climate.

s) After a minor reduction in algal biomass, an algal community develops which is adapted to being mixed. Large unicellular or filamentous algal forms appear. Among them diatoms become increasingly important with the progress of autumn.

t) This association of poorly-ingestible algae is accompanied by a variable biomass of small, edible algae.

u) This algal composition together with some reduction in fish predation pressure leads to an autumnal maximum of zooplankton which includes larger forms and species.

v) A reduction of light energy input results in a low or negative net primary production and an imbalance with the algal losses, which causes a decline of algal biomass to the winter minimum.

w) Herbivore biomass decreases as a result of reduced fecundity due both to lower food concentrations and to decreasing temperature.

x) Some species in the zooplankton produce resting stages at this time, whereas other species produced resting stages earlier.

y) At this period in the year, some cyclopoid species 'awake' from their diapauses and contribute to the over-wintering populations in the zooplankton.

Reference Cited

Sommer, U., Gliwicz, Z.M., Lampert, W., and Duncan, A., 1986, The PEG-model of seasonal succession of planktonic events in fresh waters: Archiv für Hydrobiologie, v. 106, no. 4, p. 433–471.

www.ingramcontent.com/pod-product-compliance
Lightning Source LLC
Chambersburg PA
CBHW080255180526
45167CB00006B/2536